Mathematical Surveys and Monographs

Volume 232

Linear Holomorphic Partial Differential Equations and Classical Potential Theory

Dmitry Khavinson
Erik Lundberg

American Mathematical Society
Providence, Rhode Island

2010 *Mathematics Subject Classification*. Primary 35A20, 31B20, 32A05, 30B40, 14P05.

For additional information and updates on this book, visit
www.ams.org/bookpages/surv-232

Library of Congress Cataloging-in-Publication Data

Names: Khavinson, Dmitry, 1956– author. | Lundberg, Erik, 1983– author.
Title: Linear holomorphic partial differential equations and classical potential theory / Dmitry Khavinson, Erik Lundberg.
Description: Providence, Rhode Island: American Mathematical Society, [2018] | Series: Mathematical surveys and monographs; volume 232 | Includes bibliographical references and index.
Identifiers: LCCN 2017055519 | ISBN 9781470437800 (alk. paper)
Subjects: LCSH: Potential theory (Mathematics) | Differential equations, Linear. | Differential equations, Partial. | Holomorphic functions. | AMS: Partial differential equations – General topics – Analytic methods, singularities. msc | Potential theory – Higher-dimensional theory – Boundary value and inverse problems. msc | Several complex variables and analytic spaces – Holomorphic functions of several complex variables – Power series, series of functions. msc | Functions of a complex variable – Series expansions – Analytic continuation. msc | Algebraic geometry – Real algebraic and real analytic geometry – Real algebraic sets. msc
Classification: LCC QA404.7 .K43 2018 | DDC 515/.9–dc23
LC record available at https://lccn.loc.gov/2017055519

10 9 8 7 6 5 4 3 2 1 23 22 21 20 19 18

Dedicated to Harold S. Shapiro on the occasion of his 90th birthday

Contents

Preface

-Why do solutions of linear analytic PDE suddenly break down, while for ODE the solutions are as good as the coefficients? What is the source of these mysterious singularities, and how do they propagate?

-Is there a mean value property for harmonic functions in ellipsoids similar to the well-known mean value property for balls? What about other domains with algebraic boundaries?

-What makes the integrals of all summable harmonic functions over the non-convex complement of a parabola vanish but this "null mean value property" fails for hyperbolas?

-Is there a reflection principle for harmonic functions in higher dimensions similar to the celebrated Schwarz reflection in the plane?

-Given a series of zonal harmonics, is it possible to pinpoint where the singularities occur on the boundary of its ball of convergence?

-How can we understand a moving interface between two fluids, one of them viscous and one inviscid, and what does this nonlinear dynamical problem have to do with the static inverse problem of linear potential theory?

-Why does the logarithmic potential of a uniformly charged disk experience a logarithmic singularity at the center, while that of an ellipse has square-root-type algebraic singularities at the foci?

-How far outside of their natural domain can solutions of the classical Dirichlet problem be extended? Where do the continued solutions eventually break down and why?

This book is intended as an invitation for graduate students and young analysts to these and many other intriguing questions. In trying to make the book as accessible as possible to a wide audience, we do not assume that the readers are experts in the theory of holomorphic partial differential equations. Instead we have tried to develop all necessary tools to give a good first taste of a subject rich with deep and beautiful results that illustrate a nice interplay between various parts of modern analysis and attractive themes in classical "physical" mathematics of the nineteenth century. We hope that most of the book is accessible to anyone familiar with multivariate calculus and some basics in complex analysis and differential equations.

At no point at all have we tried to produce an encyclopedic treatise, so whenever a choice between clarity and simplicity vs. generality appeared we have most decidedly chosen the former but supply enough references to satisfy an engaged and more demanding reader. (Classic PDE books by J. Hadamard and R. Courant and D. Hilbert, and later by P. Garabedian, F. John, or a more recent treatise by L. Hormander can serve as excellent supplements for this volume.) Along the same line, wherever possible, we have tried to maintain an informal way of communicating with the reader following the "Socratic method" of a dialogue rather than a formal style of a textbook.

Although the book is not intended as a regular PDE textbook, one might adopt it as a text for a graduate topics course in holomorphic PDE. It is then intended that Chapters 1–8 be covered in order; after that, there is a lot of flexibility for the instructor to build his/her own subset of the remaining chapters. We have added some exercises to most chapters in order to simplify the task of building such a topics course. Concluding each chapter, we have included Notes that supply additional references. We also point out (a great many) tantalizing open problems that could tempt a young researcher.

The second part of the book deals with deeper, more recent results and, accordingly with our plan to keep the book accessible to students, we often resort to only treating the simplest possible cases of the results discussed there, hopefully providing enough guidance and references for the readers to do deeper research on their own.

The book has grown out of efforts in giving short and long courses on the subject over the years at conferences and as topics courses. The first author started research on these topics jointly with Professor Harold S. Shapiro from the Royal Institute of Technology in Stockholm almost three decades ago. Harold's influence on the choice of material and style is far beyond what one may see from the references to his works. Of course, he bears no responsibility for any possible errors.

The first author would also like to use this opportunity and express his gratitude to his father, the late S. Ya. Khavinson, Professor J. Wermer (Brown University), and Professor P. L. Duren (University of Michigan) for many years of guidance, friendship, and support.

The second author would like to thank A. Lerario (SISSA) for a close friendship and many dynamic collaborations.

Over the years we have benefited greatly from numerous stimulating discussions on a number of related topics with Professors B. Gustafsson, G. Johnsson, and H. Shahgholian of the Royal Institute of Technology, Professor Ebenfelt from UCSD at La Jolla, Professor L. Karp at Carmiel College in Galilee, Professors S. Gardiner and H. Render from University College Dublin, Professor R. Teodorescu at the University of South Florida, Professor M. Putinar at UCSB, and Professors S. R. Bell and A. Eremenko from Purdue University. It is our great pleasure to thank them here. We hope that the book will serve as a small token of our gratitude and appreciation to them all for sharing their time and ideas with us.

Acknowledgments

The first author gratefully acknowledges support over the years by grants from the NSF and the Simons Foundation. The second author acknowledges support from Florida Atlantic University.

CHAPTER 1

Introduction: Some Motivating Questions

"The shortest and best way between two truths of the real domain often passes through the imaginary one."

- P. Painleve - J. Hadamard

1. Continuation of potentials

In the 1914 treatise [**83**] G. Herglotz studied the continuation of potentials inside the region occupied by masses. Suppose $\Omega \subset \mathbb{R}^3$ is a bounded domain with a real analytic boundary, and $p(x)$ is a polynomial mass density, where $x = (x_1, x_2, x_3)$. (The case of uniform density $p(x) = 1$ is especially interesting.) Consider the Newtonian potential

$$U_{\Omega,p} = \frac{-1}{4\pi} \int_\Omega \frac{p(y)dy}{|x-y|},$$

which is a harmonic function for $x \notin \Omega$.

Question A (G. Herglotz, 1914): How far can $U_{\Omega,p}$ be analytically continued across $\partial\Omega$ as a harmonic function inside Ω (into the region occupied by mass, or charge)?

EXAMPLE 1.1 (a familiar example). $\Omega := \{x \in \mathbb{R}^n : |x| < 1\}$ is a ball, and density $p(x) = 1$ is constant. Then the Newtonian potential

$$U_{\Omega,p}(x) = \frac{-1}{4\pi} \int_\Omega \frac{dy}{|x-y|} = \frac{C}{|x|}, \tag{1.1}$$

with C a constant, so $U_{\Omega,p}(x)$ extends into $\Omega \setminus \{0\}$. (This follows from the mean-value property for harmonic functions.) The right hand side of (1.1) is harmonic in $\mathbb{R}^3 \setminus \{0\}$. So $U_{\Omega,p}(x)$ extends to $\mathbb{R}^3 \setminus \{0\}$.

It is less well-known (and we shall prove it later in Chapter 14) that the same holds for any polynomial (or, even, entire) density p on the ball–the potential $U_{\Omega,p}(x)$ extends harmonically to $\mathbb{R}^3 \setminus \{0\}$.

As suggested in this familiar example, the problem can be stated in more physical terms:

Consider the exterior gravitational potential generated by a uniform massive object. Find a gravi-potentially equivalent object (of perhaps non-uniform density) occupying a smaller interior region.

EXAMPLE 1.2 (oblate spheroid–e.g., a planet with an "equatorial bulge"). Consider

$$\Omega := \left\{ x : \frac{x_1^2}{a^2} + \frac{x_2^2}{a^2} + \frac{x_3^2}{b^2} - 1 < 0 \right\},$$

with $a > b > 0$.

Then $U_{\Omega,p}(x)$ extends into $\Omega \setminus \{(x_1, x_2, 0) : x_1^2 + x_2^2 \le a^2 - b^2\}$. Moreover, $U_{\Omega,p}(x)$ extends as a multi-valued algebraic function across the interior of the disk.

EXAMPLE 1.3 (prolate spheroid). Consider

$$\Omega := \left\{ x : \frac{x_1^2}{a^2} + \frac{x_2^2}{b^2} + \frac{x_3^2}{b^2} - 1 < 0 \right\},$$

with $a > b > 0$.

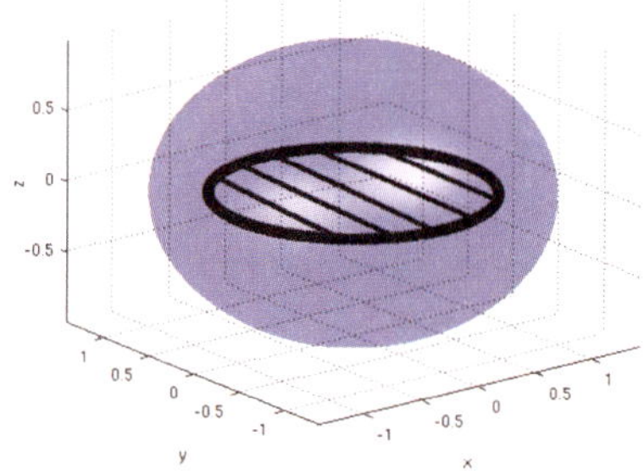

FIGURE 1. An oblate spheroid.

Then $U_{\Omega,p}(x)$ extends into $\Omega \setminus \{(x_1, 0, 0) : |x_1| \le \sqrt{a^2 - b^2}\}$, yet singularities are worse, of logarithmic type.

One of the goals of this book is to explain these examples in a unified way.

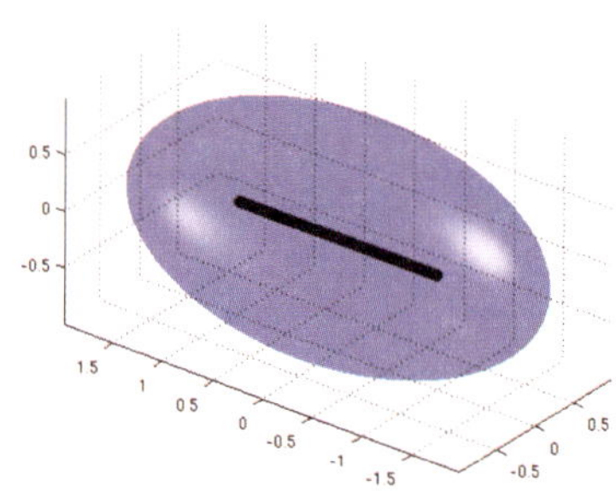

FIGURE 2. A prolate spheroid.

2. Uniqueness of potentials

Consider the spherical shell $\Omega := \{x \in \mathbb{R}^3 : r < |x| < R\}$, and let u be harmonic in Ω and vanish on the segment $(-R, -r)$ of, say, the x_1-axis.

Question B: Does u also vanish on the remaining part of the x_1-axis, the segment (r, R)?

EXERCISE 1.1. Answer the same question in $\mathbb{R}^2$ (Hint: Use the Schwarz reflection principle). Does it matter that the line passes through the center. What happens if it doesn't?

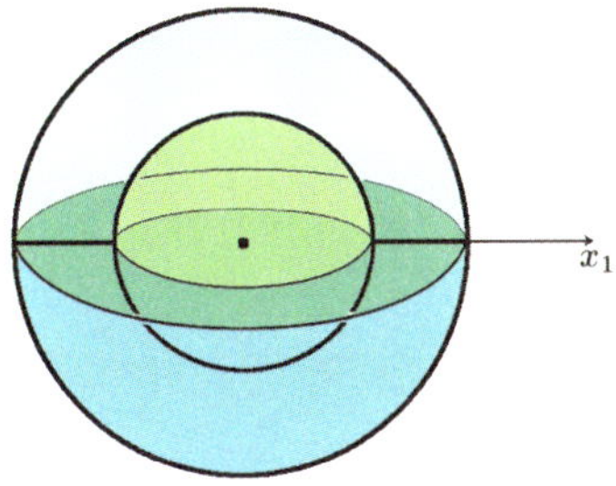

FIGURE 3. Ω is a spherical shell.

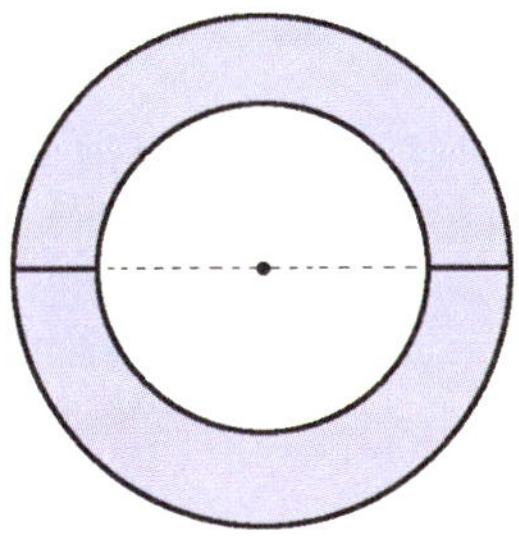

FIGURE 4. Exercise: Show that for $n = 2$ the answer is "Yes".

3. The Schwarz reflection principle

Recall [**176**, Sec. 1.2] that if γ is a real analytic curve in $\mathbb{R}^2$, then for any point A sufficiently close to γ, there exists B on the other side of γ such that $u(A) + u(B) = 0$ for all functions harmonic near γ and vanishing on γ. (The simplest cases being (i) when γ equals the real axis with $B = \overline{A}$ and (ii) when γ is the unit circle with $B = 1/\overline{A}$.)

Question C: What happens in $\mathbb{R}^n$ for $n \geq 3$?

If $\gamma = \{x_n = 0\}$ is a hyperplane, the reflection principle holds with $A = (x_1, ..., x_n)$, $B = (x_1, ..., -x_n)$ the reflected point.

For spheres, e.g., $\gamma = \{|x| = 1\}$, the principle holds in the more complicated form

$$u(x) + |x|^{2-n} u\left(\frac{x}{|x|^2}\right).$$

So, $B = \frac{A}{|A|^2}$, but there is a coefficient $|A|^{2-n}$. This is known as Kelvin's reflection. What happens in $\mathbb{R}^n$ for other surfaces? It turns out that the answer depends on the parity of n [**40**], and our goal is to explain why.

4. Szegő's theorem

Let $u = \sum_{n=0}^{\infty} a_n r^n P_n(\cos\theta)$ be an axially symmetric harmonic function in the open unit ball, $a_n \in \mathbb{R}$, $\overline{\lim}|a_n|^{1/n} = 1$, P_n are Legendre polynomials [**99**].

Since the expansion obviously diverges for $r > 1$ (it is a good exercise in classical analysis to check it.–cf. [**186**]) u must have singularities on the unit sphere. The question is, "where?" The remarkable theorem of Szegő pinpoints them precisely.

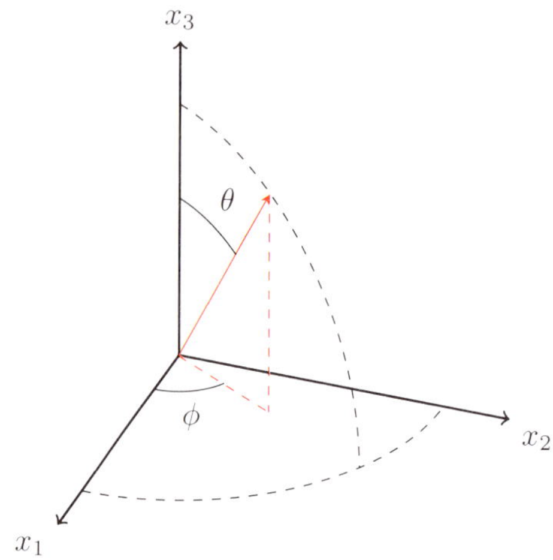

FIGURE 5. Spherical coordinates.

THEOREM 1.1 ([**186**]). *The function* $u = \sum_{n=0}^{\infty} a_n r^n P_n(\cos\theta)$ *extends harmonically across* $p_0 = (1, \theta_0, \phi)$ *if and only if the Taylor series* $f(\zeta) = \sum_{n=0}^{\infty} a_n \zeta^n$ *extends holomorphically across* $\zeta_0 = e^{i\theta_0}$.

It turns out that the two expansions appearing in the statement of Szegő's theorem can be viewed as solutions to a Cauchy problem with the same data but for different PDE (in two variables). This leads us to ask the following.
Question D: Can Szegő's theorem be understood from the high ground point of view of analytic continuation of solutions to holomorphic PDE?

5. PDE vs. ODE

Recall that for linear ODE

$$w^{(n)} + a_{n-1}(z)w^{(n-1)} + .. + a_0 w(z) = f(z),$$

where a_j, f are assumed to be analytic in a domain $\Omega \subset \mathbb{C}$, all solutions of the Cauchy problem (assume $0 \in \Omega$)

$$w(0) = w_0, w'(0) = w_1, .., w^{(n-1)}(0) = w_{n-1},$$

for any data $w_0, .., w_{n-1}$ extend analytically to all of Ω (solutions are as good as the coefficients [**90**]).

For PDE this fails miserably.

Example: Consider in $\mathbb{C}^2$ the Cauchy problem

$$\frac{\partial w}{\partial z_2} = z_1^2 \frac{\partial w}{\partial z_1}, \quad w(z_1, 0) = z_1.$$

The coefficients are analytic in all of $\mathbb{C}^2$, yet the solution

$$w(z_1, z_2) = \frac{z_1}{1 - z_1 z_2}$$

blows up arbitrarily close to the initial curve $\{(z_1, 0)\}$.

Question E: Where do these singularities come from, and what positive statements can we make in general about the size of the domain of analyticity for solutions to PDE with holomorphic coefficients?

6. Laplacian growth and the inverse potential problem

A one-parameter family $\Omega_t \subset \mathbb{R}^n$ of smoothly-varying domains each with smooth boundary is referred to as a *Hele-Shaw flow with source (or sink) at* x_0 if the normal velocity of the moving boundary $\partial\Omega_t$ coincides (at each boundary point at each time) with the normal derivative of the classical harmonic Green function of Ω_t with singularity at x_0.

This nonlinear dynamical system arises as a simplified model for the growth of a region of viscous fluid in a porous medium with injection at x_0. In the case when Ω_t is unbounded with a sink located at infinity, the process is referred to as *Laplacian growth* and arises (at least as a first-approximation) in many physical processes involving a moving-boundary, such as viscous fingering, electrodeposition, formation of ice in the presence of undercooling, growth of mineral dendrites, growth of certain types of cancerous tumors, and generally diffusion-limited growth.

S. Richardson [**160**] discovered infinitely many conservation laws for this system and established that the problem of finding the shape of Ω_T at some final time $t = T$ (given the initial shape Ω_0) is equivalent to the inverse potential problem, that is, it reduces to answering the following question.

Question F: What is the shape of a uniformly massive object given the Newtonian gravitational potential it creates outside of itself?

7. Some basic notation

Here we list some basic notation. Further notations will be introduced later in the text as needed.

Let $z \in \mathbb{C}^n$, $z = (z_1, z_2, .., z_n) = x + iy$, where $x, y \in \mathbb{R}^n$. Then,

$$\langle z, \xi\rangle := \sum_{j=1}^{n} z_j \bar{\xi}_j,$$

$$z \cdot \xi := \sum_{j=1}^{n} z_j \xi_j,$$

$$\|z\| := \langle z, z\rangle^{1/2}.$$

A multi-index α is a vector $(\alpha_1, \alpha_2, .., \alpha_n)$, with $\alpha_j \in \mathbb{Z}_+$ for $j = 1, 2, .., n$. We set

$$\begin{aligned}
|\alpha| &:= \alpha_1 + \alpha_2 + .. + \alpha_n,\\
\alpha! &:= \alpha_1!\alpha_2! \cdots \alpha_n!,\\
z^\alpha &:= z_1^{\alpha_1} z_2^{\alpha_2} \cdots z_n^{\alpha_n},\\
\partial_j &:= \frac{\partial}{\partial z_j},\\
\partial^\alpha &:= \left(\frac{\partial}{\partial z_1}\right)^{\alpha_1} \left(\frac{\partial}{\partial z_2}\right)^{\alpha_2} \cdots \left(\frac{\partial}{\partial z_n}\right)^{\alpha_n}.
\end{aligned}$$

We will sometimes use D^α in place of ∂^α.

We say that a function $f(z)$ is holomorphic near z^0 if

$$f(z) = \sum_{|a|=0}^{\infty} \frac{\partial^\alpha f(z^0)}{\alpha!}(z - z^0)^\alpha,$$

and the series converges absolutely and uniformly in some neighborhood of z^0.

The polydisk $D(z^0, R)$ of radius $R > 0$ centered at z^0 is the set

$$D(z^0, R) := \left\{ z : \left| z_j - z_j^0 \right| < R, j = 1, 2, .., n \right\}.$$

Finally, we sometimes use $z' = (z_1, z_2, .., z_{n-1})$ to abbreviate the projection of z onto the hyperplane $z_n = 0$.

Notes

The classical monographs of O. Kellogg [**99**] and G. Evans [**50**] or later ones by N. Landkof [**120**] and T. Ransford [**153**] could serve as excellent references for potential theory in depth. The rudiments of elementary several complex variables can be found in B.V. Shabat [**171**] or L. Hormander [**87**].

The problem discussed in Section 1 is stated in the Herglotz memoir [**83**], although there are earlier works of C. Neumann, E. Schmidt, and others. A fairly detailed survey of the literature can be found in [**110**] and even more so in [**176**]. Herglotz has completely solved the problem in two dimensions; in higher dimensions it remains unsolved even today (cf. [**110, 176**]). Examples (1.2) and (1.3) are studied in great detail in G. Johnsson's thesis [**94**], cf. Chapters 19 and 20. The question in Section 2 was posed and answered in the affirmative by N. Nadirashvili [**144**] by making use of symmetrization techniques. At the end of Chapter 9 we shall outline a different approach to it, which will allow us to consider more general, not necessarily symmetric, configurations. We return to the discussion of Question E in Section 5 in Chapters 18 - 20. [**176**] contains a detailed discussion of various topics associated with the Schwarz Reflection Principle and gives detailed references. A complete solution of the problem posed in Section 3 has been obtained in [**40**]. In Chapter 21, we will return to the fluid dynamic problem stated in Section 6 while giving special attention to quadrature domains, an important class of exact solutions.

CHAPTER 2

The Cauchy-Kovalevskaya Theorem with Estimates

THEOREM 2.1. *Let $\beta = (0, \ldots, m)$ and suppose that for each $\alpha \in \mathbb{Z}_+^n$, $|\alpha| \leq m$, $\alpha \neq \beta$ the functions $a_\alpha(z)$ and $f(z)$ are holomorphic in $\overline{D(0,R)}$. Then given $t < 1$, there exists a unique holomorphic solution $w = w(z)$ to the problem*

$$\begin{cases} \partial^\beta w = \displaystyle\sum_{|\alpha| \leq m, \alpha \neq \beta} a_\alpha(z) \partial^\alpha w + f(z), \\ \partial_n^j w\big|_{\{z_n = 0\}} = 0, \quad 0 \leq j \leq m-1, \end{cases} \tag{2.1}$$

defined in the polydisk $\{z : |z_k| < tR, k = 1, 2, .., n-1, |z_n| < \delta t R\}$. Most importantly, the quantity $\delta > 0$ only depends on t and the size of the coefficients $a_\alpha(z)$ but does not depend on $f(z)$.

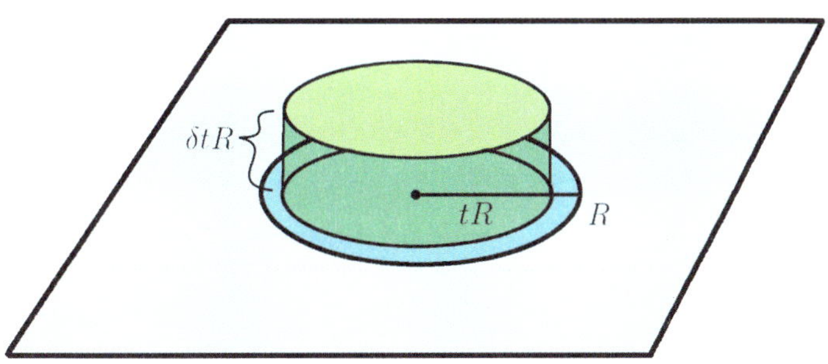

FIGURE 1. The scaling $\delta > 0$ does not depend on $f(z)$.

1. Proof of uniqueness

PROOF. *Uniqueness.* If $w_1(z)$, $w_2(z)$ were two solutions then $w = w_1 - w_2$ would satisfy the *homogeneous* equation

$$\begin{cases} \partial^\beta w = \displaystyle\sum_{|\alpha| \leq m, \alpha \neq \beta} a_\alpha(z) \partial^\alpha w, \\ \partial_n^j w\big|_{\{z_n = 0\}} = 0, \quad 0 \leq j \leq m-1. \end{cases} \tag{2.2}$$

Starting with the initial conditions in (2.2) and further differentiating in the z'-directions, we have $\partial^\alpha w(z', 0) = 0$ for any α with $\alpha_n \leq m-1$. Then, restricting the PDE in (2.2) to $\{z_n = 0\}$, we obtain

$$\partial_n^m w(z', 0) = \sum_{|\alpha| \leq m, \alpha \neq \beta} a_\alpha \partial^\alpha w(z', 0) = 0. \tag{2.3}$$

Further differentiating (2.3) in the z'-directions, we have $\partial^\gamma \partial_n^m w(z',0) = 0$ for any multi-index γ with $\gamma_n = 0$. In other words, $\partial^\alpha w(z',0) = 0$ for each α with $\alpha_n \le m$. Differentiating both sides of the homogeneous PDE (2.2) with respect to z_n we obtain $\partial_n^{m+1} w$ expressed as a combination of partial derivatives $\partial^\alpha w$ with $\alpha_n \le m$. Thus, $\partial_n^{m+1} w(z',0) = 0$ as well, and in the same manner as above this leads to $\partial^\alpha w(z',0) = 0$ for all α with $\alpha_n \le m+1$. Proceeding in this way leads to $\partial^\alpha w(0,0) = 0$ for all multi-indices α, and this implies that $w(z) \equiv 0$ by the uniqueness property of holomorphic functions. Thus $w_1(z) = w_2(z)$. $\square$

2. Proof of existence

Existence. The proof of existence shall rely upon the following proposition.

PROPOSITION 2.1. *Let $A_\alpha(z)$, $F(z)$ be holomorphic in the polydisk $\overline{D(0,1)}$. Then, given $t : 0 < t < 1$ there exists a unique holomorphic solution $w(z)$ in $D(0,t) = \{z : |z_i| < t, i = 1,..,n\}$ to the problem*

$$\begin{cases} \partial^\beta w = \displaystyle\sum_{|\alpha| \le m, \alpha \ne \beta} A_\alpha(z) \partial^\alpha w + F(z), & \beta = (0,\dots,m), \\ \partial^\alpha w(z',0) = 0, \quad |\alpha| \le m-1, \end{cases} \tag{2.4}$$

provided that

$$\sup_{z \in D(0,1)} \sum_{|\alpha| \le m, \alpha \ne \beta} |A_\alpha(z)| \le C(t),$$

where $C(t)$ is a sufficiently small constant that depends only on t.

Before proving the proposition, let us continue with the proof of the C-K theorem. So, let $0 < t < 1$ and choose $\delta > 0$ such that $\{|z'| < R, |z_n| < \delta R\} \subset D(0,R)$. By changing variables $\xi' = \frac{z'}{R}$, $\xi_n = \frac{z_n}{\delta R}$ we arrive at the equation

$$\partial_{\xi_n}^m w(z(\xi)) = \sum_{|\alpha| \le m, \alpha \ne \beta} \delta^{m-\alpha_n} R^{m-|\alpha|} a_\alpha(z(\xi)) \partial_\xi^\alpha w(z(\xi)) + (\delta R)^m f(z(\xi))$$

or

$$\partial_{\xi_n}^m w = \sum_{|\alpha| \le m, \alpha \ne \beta} A_\alpha(\xi) \partial_\xi^\alpha w + F(\xi), \quad \xi \in D(0,1),$$

where $A_\alpha(\xi) := \delta^{m-\alpha_n} R^{m-|\alpha|} a_\alpha(z(\xi))$, $\alpha \ne \beta$. Since $\alpha \ne \beta$ implies $\alpha_n < m$, we have

$$\sup_{D(0,1)} \sum_{|\alpha| \le m, \alpha \ne \beta} |A_\alpha| = O(\delta).$$

Therefore, for a fixed $t : 0 < t < 1$, this supremum can be made smaller than the $C(t)$ stated in Proposition 2.1 provided that δ is sufficiently small, and the theorem follows.

PROOF OF THE PROPOSITION. For each k define w_{k+1} to be the solution of the Cauchy problem

$$\begin{cases} \partial^\beta w_{k+1} = \displaystyle\sum_{|\alpha| \le m, \alpha \ne \beta} A_\alpha(z) \partial^\alpha w_k + F(z), \\ \dfrac{\partial^j w_{k+1}}{\partial z_n^j} = 0, \quad 0 \le j \le m-1, \end{cases} \tag{2.5}$$

and set $w_0(z) = 0$. Existence of solutions w_k is proven below in Lemma 2.3. Now, set $v_k = w_k - w_{k-1}$, $k \geq 1$. Then for each k we have

$$\begin{cases} \partial^\beta v_{k+1} = \displaystyle\sum_{|\alpha|\leq m, \alpha\neq\beta} A_\alpha(z)\partial^\alpha v_k, \\ \dfrac{\partial^j v_{k+1}}{\partial z_n^j} = 0, \quad 0 \leq j \leq m-1. \end{cases}$$

Our immediate objective is to show the convergence of the telescopic series $\sum v_k$. This requires the following two lemmas that are simple and useful results from complex analysis in one variable that we will prove in Section 3 below.

LEMMA 2.1. *If $h(z)$ is analytic in $\mathbb{D} = \{z : |z| < 1\} \subset \mathbb{C}$, $h(0) = 0$ and*

$$|h'(z)| \leq \frac{1}{(1-|z|)^p} \quad \textit{for some } p > 0,$$

then

$$|h(z)| \leq \frac{1}{p(1-|z|)^p}.$$

LEMMA 2.2. *If $h(z)$ is analytic in $\mathbb{D}$ and*

$$|h(z)| \leq \frac{1}{(1-|z|)^p} \quad \textit{for some } p > 0,$$

then

$$|h'(z)| \leq \frac{(p+1)e}{(1-|z|)^{p+1}}.$$

As a consequence of Lemmas 2.1 and 2.2 we claim that the following estimate holds for $v_{k+1}(z)$

$$|v_{k+1}(z)| \leq \frac{N\,(Ce^m)^k}{[(1-|z_1|)\cdots(1-|z_n|)]^{mk}}, \tag{2.6}$$

where $N = N(F)$ is a constant that only depends on F and

$$C := \sup_{z\in D(0,1)} \sum_{|\alpha|\leq m,\alpha\neq\beta} |A_\alpha(z)|.$$

Assuming the claim, we have for each $z \in D(0,t)$

$$\frac{Ce^m}{[(1-|z_1|)\cdots(1-|z_n|)]^m} \leq \frac{Ce^m}{(1-t)^{nm}}$$

and provided that

$$C < C(t) := \frac{(1-t)^{nm}}{e^m},$$

the series $\sum |v_k(z)|$ is majorized by a geometric series. Hence, the series $\sum v_k$ converges uniformly inside $D(0,t)$ to a holomorphic function $w(z)$. So its partial sums w_k converge to w and, in view of (2.5), w is a solution of (2.4). □

Let us now prove the estimate (2.6). We proceed by induction. Since $w_0 = 0$ we have $v_1 = w_1$, and the estimate of $|w_1|$ by a constant $N = N(F)$, i.e., the validity of (2.6) for $k = 0$, follows from the following lemma.

LEMMA 2.3. *Suppose that $F = F(z)$ is holomorphic in the polydisk $D(0,1)$. Then the problem*

$$\left\{ \begin{array}{l} \partial^\beta w = F, \quad \beta = (0, \ldots, m), \\ \partial^j_{z_n} w\,(z', 0) = 0, \quad 0 \le j \le m-1 \end{array} \right. \tag{2.7}$$

admits a unique holomorphic solution $w = w(z)$ defined in $D(0,1)$ and satisfying $|w(z)| \le N$ on $D(0,1)$ for a positive constant N that only depends on F.

PROOF OF LEMMA 2.3. Set

$$w\,(z', z_n) = \int_0^{z_n} \frac{(z_n - \lambda)^{m-1}}{(m-1)!} F\,(z', \lambda)\, d\lambda.$$

A routine check shows that w is a solution of (2.7) and the lemma follows. □

Continuing now the proof of (2.6), assume that (2.6) holds for $k+1$, i.e.,

$$|v_{k+1}(z)| \le \frac{N\,(Ce^m)^k}{[(1-|z_1|)\cdots(1-|z_n|)]^{mk}}.$$

For $z \in D(0,1)$ we have

$$\left|\partial^\beta v_{k+2}(z)\right| \le C \sup_{|\alpha| \le m, \alpha \ne \beta} |\partial^\alpha v_{k+1}(z)|\,.$$

A repeated application of Lemma 2.2 and the inductive hypothesis then yield

$$|\partial^\alpha v_{k+1}(z)| \le \frac{\gamma(\alpha) N C e^m\,(Ce^m)^k}{[(1-|z_1|)\cdots(1-|z_n|)]^{mk}\,(1-|z_1|)^{\alpha_1}\cdots(1-|z_n|)^{\alpha_n}},$$

where

$$\gamma(\alpha) := ((mk+\alpha_1)\cdots(mk+1))\cdots((mk+\alpha_n)\cdots(mk+1))\,.$$

Thus, taking into account that for each $z \in D(0,1)$, and $\alpha \in \mathbb{Z}^n_+$, $|\alpha| \le m$

$$(1-|z_1|)^{\alpha_1}\cdots(1-|z_n|)^{\alpha_n} \ge (1-|z_1|)^m \cdots (1-|z_n|)^m\,,$$

and applying repeatedly Lemma 2.1 we obtain

$$|v_{k+2}(z)| \le \frac{\gamma(\alpha)}{[m(k+1)]^m}\,\frac{N\,(Ce^m)^{k+1}}{[(1-|z_1|)]^{m(k+1)}}$$

and (2.6) follows, for obviously,

$$\frac{\gamma(\alpha)}{[m(k+1)]^m} \le 1.$$

3. Proofs of accessory lemmas: Fun and useful inequalities

Finally, let us supply the proofs of Lemmas 2.1 and 2.2; the proof of the former is simply based on estimating the integral provided by the fundamental theorem of calculus, and the proof of the latter is based on estimating the integral provided by the Cauchy integral formula.

PROOF OF LEMMA 2.1. For each $z \in \mathbb{D}$ we have

$$h(z) = \int_0^z h'(\xi)\, d\xi.$$

Thus,

$$|h(z)| \le \int_0^{|z|} \frac{dt}{(1-t)^p} \le \frac{1}{p(1-|z|)^p},$$

for $p: 0 < p \le 1$. If $p > 1$ we must check that

$$\frac{1}{p-1}\left((1-|z|)^{1-p} - 1\right) \le \frac{1}{p(1-|z|)^p},$$

or, setting $\zeta = 1 - |z|$,

$$\zeta\left(1 - \zeta^{p-1}\right) \le \frac{p-1}{p}, \quad 0 \le \zeta \le 1.$$

But the maximum value of the expression in the left hand side is attained at $\zeta_0 = p^{-\frac{1}{p-1}}$, that value being $p^{-\frac{1}{p-1}}\left(\frac{p-1}{p}\right)$. That number is obviously less than $\frac{p-1}{p}$ provided $p > 1$. □

Proof of Lemma 2.2. Take $z \in \mathbb{D}$. For any $s: 0 < s < 1$ the circle $\gamma := \{\xi : |\xi - z| = s(1-|z|)\}$ is contained in $\mathbb{D}$. Using Cauchy's formula we obtain

$$|h'(z)| \le \frac{1}{2\pi s(1-|z|)}\left\{\int_0^{2\pi} \left|h\left(z + s(1-|z|)e^{i\theta}\right)\right| d\theta\right\}.$$

Since

$$(1-|\xi|)^p \ge (1-|z|)^p(1-s)^p$$

for each $\xi \in \gamma$,

$$|h'(z)| \le \frac{1}{(1-|z|)^{p+1}s(1-s)^p}.$$

Let us choose $s = \frac{1}{p+1}$ (this optimizes the above estimate, since $\frac{1}{p+1}\left(1 - \frac{1}{p+1}\right)^p$ is the maximum value of $s(1-s)^p$ for $0 < s < 1$). To get the estimate in the statement of the lemma it now suffices to show that

$$\frac{1}{p+1}\left(1 - \frac{1}{p+1}\right)^p \ge \frac{1}{e(p+1)} \tag{2.8}$$

for each $p > 0$. If we put $\zeta = \frac{1}{p+1}$, then (2.8) is equivalent to showing that $\log(1-\zeta) + \frac{\zeta}{1-\zeta} > 0$ for $0 < \zeta < 1$. The latter follows, e.g., from the identity

$$\log(1-\zeta) + \frac{\zeta}{1-\zeta} = \sum_{n\ge 2} \frac{n-1}{n}\zeta^n, \quad 0 < \zeta < 1.$$

Therefore,

$$|h'(z)| \le \frac{(p+1)e}{(1-|z|)^{p+1}}$$

as desired. □

Notes

Almost every textbook on PDEs contains a statement and a proof of the C-K theorem, the only truly general theorem in the theory of PDE, although usually coached in a slightly different language of real-analytic functions (cf. e.g., [**58**, **72**, **93**]). The statement presented here, and specifically the proof based on Picard's method of iterations, often used in ODEs, are from Hörmander's book [**86**]. The main feature of Hörmander's version that makes it so crucial for applications is a rather precise estimate on the size of the domain of existence for the solution to Cauchy's problem that is independent of how large the data is, or as presented in this section how large the non-homogeneous term is, provided that they (that is, the data, or the non-homogeneous term) are holomorphic in a sufficiently large domain.

CHAPTER 3

Remarks on the Cauchy-Kovalevskaya Theorem

1. The Cauchy problem with holomorphic data

In the statement of the C-K theorem, the data were assumed to vanish. One can replace the zero data by any data holomorphic in $\overline{D(0', R)}$, and this case may be reduced to the previous by assimilating the data into the nonhomogeneous term $f(z)$. To be more explicit, suppose that in (2.1) we ask for a solution w satisfying

$$\begin{aligned} w\big|_{\{z_n=0\}} &= g_0(z'), \\ \partial_n w\big|_{\{z_n=0\}} &= g_1(z'), \\ \dots \quad & \dots \\ \partial_n^{m-1} w\big|_{\{z_n=0\}} &= g_{m-1}(z'), \end{aligned}$$

where $g_0, \dots, g_{m-1}$ are arbitrary functions of $(n-1)$ variables holomorphic in $\overline{D(0', R)} = \{z : |z_k| \leq R,\ k = 1, 2, .., n-1\}$.

Consider $G(z) := g_0(z') + \frac{z_n}{1!} g_1(z') + \dots + \frac{z_n^{m-1}}{(m-1)} g_{m-1}(z')$. Then

$$(\partial_n)^k (G - w)\Big|_{\{z_n=0\}} = 0,$$

for all $k \leq m-1$.

Thus, we have

$$\begin{aligned} \partial_n^m w &= (\partial_n)^m (w - G) \\ &= \sum_{\substack{|\alpha| \leq m \\ \alpha \neq (0,\dots,m)}} a_\alpha(z) \partial^\alpha (w - G) + f + \sum_{\substack{|\alpha| \leq m \\ \alpha \neq (0,\dots,m)}} a_\alpha(z) \partial^\alpha G. \end{aligned}$$

Since G is holomorphic in $\overline{D(0', R)} \times \mathbb{C}$,

$$f_1(z) := f(z) + \sum_{\substack{|\alpha| \leq m \\ \alpha \neq (0,\dots,m)}} a_\alpha(z) \partial^\alpha G$$

is holomorphic in $\overline{D(0, R)}$.

Since the data has been assimilated into the nonhomogeneous term $f(z)$ the final statement in the conclusion of the C-K theorem thus carries over:
The quantity $\delta > 0$ does not depend on the size of the data.

EXERCISE 3.1. Consider the Cauchy problem discussed in Section 5 of Chapter 1:

$$\frac{\partial w}{\partial z_2} = z_1^2 \frac{\partial w}{\partial z_1}, \quad w(z_1, 0) = z_1.$$

Follow the proof of the Cauchy-Kovalevskaya Theorem in order to obtain an explicit value of $\delta(R,t)$ (as a function of R and t). Letting $R \to \infty$, compare the asymptotic behavior of your answer $\delta(R,t)$ with the known singularities of the explicit solution

$$w(z_1, z_2) = \frac{z_1}{1 - z_1 z_2}.$$

EXERCISE 3.2. Determine an estimate for $\delta(R,t)$ for the Cauchy problem:

$$\begin{cases} w_{z_2 z_2} = w + z_1 z_2 w_{z_1} + 3z_1^2 z_2 w_{z_1 z_2} + z_1^3, \\ w(z_1, 0) = w_{z_2}(z_1, 0) = 0. \end{cases}$$

EXERCISE 3.3. Consider the Cauchy problem for Laplace's equation:

$$\begin{cases} w_{z_1 z_1} + w_{z_2 z_2} + .. + w_{z_n z_n} = 0, \\ w(z', 0) = g(z', 0), \\ w_{z_n}(z', 0) = g_{z_n}(z', 0). \end{cases}$$

Suppose that $g(z)$ is an entire function, that is a function holomorphic in all of $\mathbb{C}^n$, and show by direct estimates that the solution $w(z)$ is entire.

2. Transversality of the highest-order derivatives

In the case when the order of derivatives on the right hand side of the Cauchy problem is larger than m the theorem may fail altogether.

The following example which is due to Kovalevskaya herself [**119**] illustrates the above comment rather well.

Consider the "heat equation"

$$\begin{cases} \frac{\partial w}{\partial z_2} = \frac{\partial^2 w}{\partial z_1^2}, \\ w(z_1, 0) = f(z_1), \end{cases} \tag{3.1}$$

with data f being a holomorphic function near the origin.

Suppose $w(z_1, z_2)$ is a solution holomorphic in a bi-disk centered at the origin. Then we can write

$$w(z_1, z_2) = \sum_{n=0}^{\infty} a_n(z_1) z_2^n,$$

where $a_0(z_1) = f(z_1)$ and all a_n's are holomorphic in a disk $\{|z_1| < R\}$.

Applying the equation we find

$$\sum_{n=1}^{\infty} n a_n(z_1) z_2^{n-1} = \sum_{n=1}^{\infty} a''_{n-1} z_2^{n-1}.$$

Hence $n a_n(z_1) = a''_{n-1}$ for all $n \geq 1$. Since $a_0(z_1) = f(z_1)$ we obtain that $a_n(z_1) = \frac{f^{(2n)}(z_1)}{n!}$.

In particular, $w(0, z_2) = \sum_{n=0}^{\infty} \frac{f^{(2n)}(0)}{n!} z_2^n$. So, if we take, e.g., $f(z_1) := \sum z_1^n$, we have $f^{(2n)}(0) = (2n)!$ and that gives us the expansion $w(0, z_2) = \sum_{n=0}^{\infty} \frac{(2n)!}{n!} z_2^n$ that converges nowhere!

Moreover, the following exercise gives a necessary condition for existence of a holomorphic solution of this problem (cf. [**114**]).

EXERCISE 3.4. Suppose $w(z_1, z_2)$ is a holomorphic solution to (3.1). Show that $f(z_1)$ must be an entire function of order at most 2, i.e., $|f(z_1)| \leq \text{const} \cdot \exp\left(C|z_1|^2\right)$.

3. The C-K theorem for non-singular hypersurfaces

Suppose Γ is a non-singular analytic hypersurface in a neighborhood U of $z^0 \in \Gamma$, i.e., Γ is the zero set

$$\Gamma = \{z \in \mathbb{C}^n : \varphi(z) = 0\},$$

of a function φ holomorphic in U, with $\nabla_z \varphi := (\partial_1 \varphi, \ldots, \partial_n \varphi) \neq (0, \ldots, 0)$ in U.

We want to solve the following Cauchy problem

$$\begin{cases} \sum\limits_{|\alpha|=m} a_\alpha(z)\partial^\alpha w = f(z); \\ \partial^\alpha(w - g) = 0 \text{ on } \Gamma \cap U, |\alpha| \leq m - 1, \end{cases} \tag{3.2}$$

where a_α, f and g are, say, holomorphic in a slightly larger neighborhood U' of z^0.

Without loss of generality we can assume $z^0 = 0$ and $\frac{\partial \varphi}{\partial z_n} \neq 0$ in perhaps slightly smaller neighborhood $V \subset U$. Then, the change of variables

$$z = (z', z_n) \to \xi := T(z) = (\xi', \xi_n),$$

given by $\xi' = z'$ and $\xi_n = \varphi(z)$ is bi-holomorphic near z^0. Change coordinates to ξ by setting $a_\alpha(z) = a_\alpha\left(T^{-1}(\xi)\right) =: A_\alpha(\xi)$, $f(z) = F(\xi)$, $g(z) = G(\xi)$, and $w(z) = W(\xi) = W(z', \varphi(z))$.

EXERCISE 3.5. With this change of coordinates, show that the differential equation in (3.2) is transformed into

$$C(\xi)\frac{\partial^m}{\partial \xi_n^m} W + \sum_{\substack{|\alpha| \leq m \\ \alpha \neq (0,\ldots,m)}} B_\alpha(\xi)\partial^\alpha W = F(\xi),$$

where the coefficients $B_\alpha(\xi)$ are composed from A_α's and the derivatives of φ, while the coefficient $C(\xi)$ is given by

$$C(\xi) = C(T(z)) := c(z) = \sum_{|\alpha|=m} a_\alpha(z)\left(\nabla_z \varphi\right)^\alpha.$$

Warning: This exercise is straightforward and satisfying, but tedious!

Thus, in order to be able to write down our Cauchy problem in ξ variables in a canonical form (cf. Chapter 2)

$$\begin{cases} \dfrac{\partial^m}{\partial \xi_n^m} W = \sum\limits_{\substack{|\alpha| \leq m \\ \alpha \neq (0,\ldots,m)}} \mathcal{D}_\alpha(\xi)\partial^\alpha W + F_1(\xi), \\ \partial^\alpha(W - G)|_{\{\xi_n = 0\}} = 0, \text{ for all } \alpha : |\alpha| \leq m - 1 \end{cases}$$

with coefficients $\mathcal{D}_\alpha$, F_1 holomorphic near the origin, it is necessary and sufficient that $c(z) \neq 0$ near z^0, or, by continuity, that

$$\sum_{|\alpha|=m} a_\alpha(z)\left(\nabla_z \varphi\right)^\alpha|_{z^0} \neq 0.$$

Thus, we have the following extension of the C-K theorem.

THEOREM 3.1. *The holomorphic solution w of the Cauchy problem*

$$\begin{cases} \sum\limits_{|\alpha|\le m} a_\alpha(z)\partial^\alpha w = f(z); \\ \partial^\alpha(w-g) = 0 \text{ on } \Gamma\cap U, \quad \text{for all } \alpha : |\alpha| \le m-1, \end{cases}$$

exists and is unique in a neighborhood V of z^0, whose size depends only on a_α, Γ, and f but not on g, provided that $c\left(z^0\right) = \sum\limits_{|\alpha|=m} a_\alpha(z)\left(\nabla\varphi\right)^\alpha|_{z^0} \neq 0$.

$c(z)$ is called the *characteristic form* of the differential operator

$$\mathcal{L} := \sum_{|\alpha|\le m} a_\alpha(z)\partial^\alpha.$$

If $c\left(z^0\right)\neq 0$, $z^0\in\Gamma$, Γ is called *non-characteristic* (w.r.t. the operator $\mathcal{L}$) at z^0.

If Γ is characteristic at z^0, i.e., $c\left(z^0\right) = 0$, both uniqueness and existence in the C-K theorem may fail.

Example 1: Recall the Cauchy problem for the heat equation

$$\begin{cases} \dfrac{\partial^2 w}{\partial z_1^2} - \dfrac{\partial w}{\partial z_2} = 0, \\ w\left(z_1, 0\right) = f\left(z_1\right), \end{cases}$$

where as we saw a solution need not exist. Since in this case

$$\Gamma = \left\{z\in\mathbb{C}^2 : \varphi(z) := z_2 = 0\right\},$$

$c(z) = \left(\frac{\partial\varphi}{\partial z_1}\right)^2 \equiv 0$, i.e., Γ is everywhere characteristic for the heat operator.

Example 2: Next we consider an example of an initial surface that is characteristic only at one point.

$\Gamma = \{z\in\mathbb{C}^2 : z_2 = f(z_1)\}$ is a graph of an analytic function with $f'(0) = 0$, e.g., $f(z_1) = z_1^3$.

If we change variables $z_1 = \frac{x+iy}{2}$, $z_2 = \frac{x-iy}{2i}$ the Laplace operator $\frac{\partial^2}{\partial x^2} + \frac{\partial^2}{\partial y^2}$ becomes $\frac{\partial^2}{\partial z_1\partial z_2}$.

Consider the Cauchy problem:

$$\begin{cases} \frac{\partial^2}{\partial z_1\partial z_2} u = 1, \text{ near } \Gamma, \\ u = 0, \nabla u = 0, \text{ on } \Gamma. \end{cases}$$

Γ is characteristic at $z^0 = (0,0)$ since $c(z^0) = -f'(0)$.

The solution is

$$u = z_1 z_2 + a(z_1) + b(z_2),$$

where a, b satisfy

$$\begin{cases} z_1 z_2 + a(z_1) + b(z_2) = 0, \text{ near } \Gamma, \\ z_2 + a'(z_1) = 0, \text{ on } \Gamma, \\ z_1 + b'(z_2) = 0, \text{ on } \Gamma. \end{cases}$$

From this it follows that $a'(z_1) = -f(z_1)$, and $b'(z_2) = -f^{-1}(z_2)$ where f^{-1} is the inverse of f and hence is singular at $z_2 = 0$.

See Fig. 1.

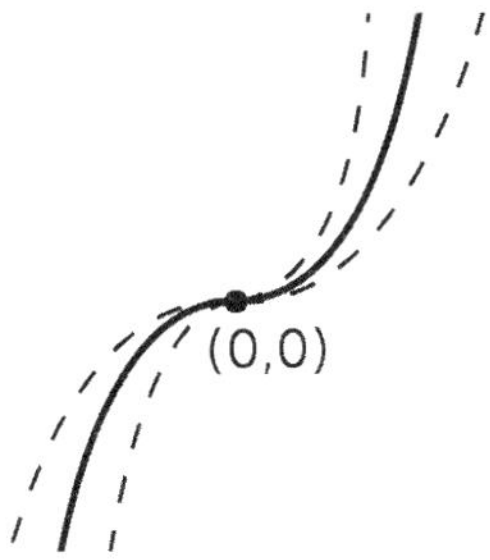

FIGURE 1. Posing the Cauchy problem near the characteristic point $(0,0)$ results in a small value of $\delta > 0$. The total region of analyticity guaranteed by the C-K theorem (taking a union of polydisks of existence) is "pinched".

Example 3: To see the failure of uniqueness consider the Laplace operator

$$\Delta := \sum_{j=1}^{n} \frac{\partial^2}{\partial z_j^2}$$

and

$$\Gamma = \left\{ z : \sum_{j=1}^{n} a_j z_j = 0 \right\}$$

with $\sum_{j=1}^{n} a_j^2 = 0$. Then Γ is everywhere characteristic w.r.t. Δ. On the other hand, for any $p \geq 2$, the function

$$w(z) = \left(\sum_{j=1}^{n} a_j z_j \right)^p$$

is harmonic in $\mathbb{C}^n$ and satisfies $w = \nabla_z w = 0$ on Γ.

4. The Goursat problem

The multi-index β in the C-K theorem need not be of the form $(0, \ldots, m)$, it can be an arbitrary multi-index $\beta = (\beta_1, \ldots, \beta_n)$ with $|\beta| = m$.

However, in that case, the data must be given in the form

$$\begin{array}{ccccccc} w\big|_{\{z_1=0\}} & = & \cdots & = & \partial_1^{\beta_1 - 1} w\Big|_{\{z_1=0\}} & = & 0 \\ . & & \cdots & . & & & . \\ . & & \cdots & . & & & . \\ w\big|_{\{z_n=0\}} & = & & = & \partial_n^{\beta_n - 1} w\big|_{\{z_n=0\}} & = & 0. \end{array}$$

This type of problem is known as the *Goursat Problem.* The statement (and the proof) of the C-K theorem for that situation remain essentially the same (cf. [**86**, §9.4]).

EXERCISE 3.6. Formulate an appropriate modification of Lemma 2.3 and Theorem 2.1 adapted to the Goursat problem.

5. Existence of the Riemann function

An important example of the Goursat problem used in the sequel is the following.

Let $a(z,w)$, $b(z,w)$, $c(z,w)$ be holomorphic in a neighborhood of a point $(\xi,\eta)\in\mathbb{C}^2$ and seek a solution R of the Cauchy problem

$$\begin{cases} \dfrac{\partial^2 R}{\partial z\partial w} = \dfrac{\partial(aR)}{\partial z} + \dfrac{\partial(bR)}{\partial w} - cR, \\ R|_{\{z=\xi\}} = \exp\left(\int_\eta^w a(\xi,\tau)d\tau\right), \\ R|_{\{w=\eta\}} = \exp\left(\int_\xi^z b(\tau,\eta)d\tau\right). \end{cases}$$

The (unique) holomorphic solution $R := R(z,w;\xi,\eta)$ whose existence follows from the general C-K theorem is called the Riemann function for the operator $\mathcal{L} : \frac{\partial^2}{\partial z\partial w} + a\frac{\partial}{\partial z} + b\frac{\partial}{\partial w} + c$ at (ξ,η) and is the key to the Riemann method of integration of the hyperbolic equations in two variables.

Notes

In presenting the C-K theorem in the most general form (Theorem 3.1) we still follow Hörmander's book [**86**]. The notion of a characteristic point together with the relevant examples associated with the heat equation appear in S. Kovalevskaya's paper [**119**]. (See [**114**] for some related historical background.) B. Riemann introduced the Riemann function in order to calculate explicitly solutions of Cauchy problems for hyperbolic equations in two variables. More on properties of the Riemann function together with other proofs of its existence and uniqueness can be found, e.g., in [**58**], [**93**], [**72**], [**81**], [**196**] and references cited therein.

CHAPTER 4

Zerner's Theorem

Suppose $u(z)$ solves a holomorphic PDE $\mathcal{L}u = f$ in Ω. Imagine that we have arrived at u not through explicit formulas, but say through the existence guaranteed by the Cauchy-Kovalevskaya Theorem. We may wish to extend $u(z)$ past Ω. To this end, one can say nothing in general (and even explicit formulas may not help), but Zerner's Theorem provides a positive result under rather mild assumptions on the domain and the PDE.

THEOREM 4.1. *Let u be a holomorphic solution of the equation $\mathcal{L}u = f$ in a domain $\Omega \subset \mathbb{C}^n$ with C^1 boundary, and assume that the coefficients a_α, $|\alpha| \leq m$ and f are holomorphic in $\bar{\Omega}$. Let $z_0 \in \partial\Omega$. If $\partial\Omega$ is (Zerner) non-characteristic at z_0 with respect to $\mathcal{L}$ then u extends holomorphically into a neighborhood of z_0.*

It is necessary to define the terminology *Zerner characteristic* appearing in the statement of the Theorem. Notice that the boundary $\partial\Omega$ is a *real* hypersurface. In the previous Chapter, the term *characteristic* referred to a property of a point on a complex hypersurface. We will define *Zerner characteristic* below, but first let us recall the basic fact that every real hyperplane contains a unique complex hyperplane.

1. Real and complex hyperplanes

Consider a differential operator with holomorphic coefficients in $\mathbb{C}^n$

$$\mathcal{L} = \sum_{|\alpha| \leq m} a_\alpha(x)\partial^\alpha,$$

and a real hyperplane H in $\mathbb{C}^n$ given by

$$H = \left\{ z : \sum_{j=1}^{n} a_j x_j + b_j y_j = t \right\},$$

where $x_j = Re\, z_j$, $y_j = Im\, z_j$ and a_j, b_j, t are real numbers. Writing $z_j = \frac{z_j + \overline{z_j}}{2}$, $y_j = \frac{z_j - \overline{z_j}}{2i}$, we can express this as

$$H = \left\{ \sum_{j=1}^{n} \frac{a_j - ib_j}{2} z_j + \frac{a_j + ib_j}{2} \overline{z_j} = t \right\}$$

or, setting $\lambda_j = a_j - ib_j$, we obtain

$$H = \left\{ Re\left(\sum_{j=1}^{n} \lambda_j z_j \right) = t \right\}.$$

Note that for a given $z^0 \in H$, the real hyperplane H contains a unique complex hyperplane Π passing through z^0, namely,

$$\Pi = \left\{ \sum_{j=1}^{n} \lambda_j z_j = t' \right\},$$

where $t' := \sum_{j=1}^{n} \lambda_j z_j^0$, so that Π indeed passes through z^0, and we see that $\Pi \subset H$ since $Re\, t' = t$.

2. Zerner characteristic hypersurfaces

DEFINITION 4.1. A real hyperplane H passing through a point $z_0 \in \mathbb{C}^n$ is called Zerner characteristic with respect to the operator $\mathcal{L}$ at z_0 if the (unique) complex hyperplane $\Pi \subset H$ with $z^0 \in \Pi$ is characteristic with respect to $\mathcal{L}$ at z_0, i.e.,

$$\sum_{|\alpha|=m} a_\alpha (z_0)\, \lambda_1^{\alpha_1} \cdots \lambda_n^{\alpha_n} = 0.$$

DEFINITION 4.2. A C^1-real hypersurface Γ given by $\{z : \varphi(z) = 0\}$, where φ is a C^1 real-valued function $\nabla\varphi \neq 0$ near a point $z_0 \in \Gamma$, is called Zerner characteristic with respect to the operator $\mathcal{L}$ at z_0 if the real hyperplane H tangent to Γ at z_0 is Zerner characteristic with respect to $\mathcal{L}$ at z_0.

Because H is the tangent plane to Γ at z_0, we can write its equation in the form

$$H : \sum_{j=1}^{n} \left.\frac{\partial \varphi}{\partial x_j}\right|_{z_0} x_j + \left.\frac{\partial \varphi}{\partial y_j}\right|_{z_0} y_j = t,$$

or,

$$H : Re \sum_{j=1}^{n} \frac{1}{2} \left. \left(\frac{\partial \varphi}{\partial x_j} - i \frac{\partial \varphi}{\partial y_j} \right) \right|_{z_0} z_j = t.$$

But then Π (the complex hyperplane tangent to Γ at z_0) is given by

$$\Pi : \sum_{j=1}^{n} \left.\frac{\partial \varphi}{\partial z_j}\right|_{z_0} z_j = t$$

where as usual

$$\frac{\partial}{\partial z_j} := \frac{1}{2} \left(\frac{\partial}{\partial x_j} - i \frac{\partial}{\partial y_j} \right),$$

and so Π is Zerner characteristic with respect to $\mathcal{L}$ at z_0 if and only if

$$\sum_{|\alpha|=m} a_\alpha (z_0) \left.(\nabla_z \varphi)^\alpha\right|_{z_0} = 0.$$

EXAMPLE 4.1. Let $\Omega = B_1(0) := \{z \in \mathbb{C}^n : \|z\| < 1\}$ and $\mathcal{L} = \sum_{j=1}^{n} \frac{\partial^2}{\partial z_j^2}$. In this case, Ω is a domain of holomorphy, i.e., there exist functions which are holomorphic in Ω and do not extend to a larger set [**171**, Ch. III, Sec. 12]. Let us find Zerner characteristic points with respect to $\mathcal{L}$ in $\partial\Omega$. The equation of $\partial\Omega$ can be written in the form

$$\sum_{j=1}^{n} z_j \overline{z_j} - 1 = 0,$$

and so $z \in \partial\Omega$ is characteristic if and only if

$$\sum_{j=1}^{n} \bar{z}_j^2 = \overline{\sum_{j=1}^{n} z_j^2} = 0.$$

Thus, except for the points that lie on the cone $\left\{ z : \sum_{j=1}^{n} z_j^2 = 0 \right\}$, called an *isotropic cone*, any solution of $\mathcal{L}u = 0$ can be continued holomorphically across all other points on the unit sphere.

3. Proof of Zerner's theorem

PROOF. Without loss of generality we may assume $z_0 = 0$, and that the normal vector $(0, 0, \ldots, 1)$ is pointing outside Ω. Thus the real hyperplane H tangent to $\partial\Omega$ at z_0 is $\{Re\, z_n = 0\}$. For $\varepsilon > 0$ consider the hyperplanes

$$H_\varepsilon : Re\, z_n = -\varepsilon.$$

By continuity, H_ε is still non-characteristic with respect to $\mathcal{L}$ if ε is sufficiently small. Let Π_ε be the unique complex hyperplane inside H_ε.

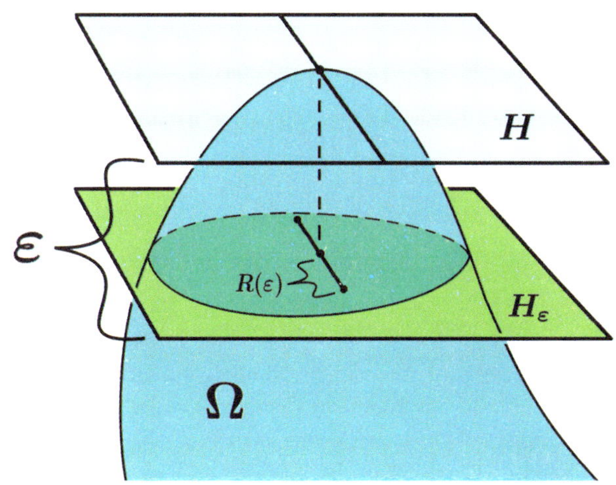

FIGURE 1. The hyperplane H_ε contains a polydisk of radius $R(\varepsilon)$ within the complex hyperplane Π_ε.

CLAIM. $\Pi_\varepsilon \cap \Omega$ contains an $(n-1)$-dimensional polydisk centered at the point $(0, \ldots, 0, -\varepsilon)$ of radius $R(\varepsilon)$ such that

$$\lim_{\varepsilon \to 0} \frac{\varepsilon}{R(\varepsilon)} = 0.$$

Assuming the claim, we have u holomorphic in the polydisk of radius $R(\varepsilon)$ inside Π_ε. Also Π_ε is non-characteristic with respect to $\mathcal{L}$. So, we can solve the Cauchy problem

$$\begin{cases} \mathcal{L}v = f & \text{near} \quad \Pi_\varepsilon \cap \Omega, \\ \partial^\alpha(v - u) = 0,\ |\alpha| \le m - 1 & \text{on} \quad \Pi_\varepsilon. \end{cases}$$

By the uniqueness for solutions of Cauchy problems, $v = u$, and hence applying the Cauchy-Kovalevskaya theorem we conclude that u is holomorphic in a polydisk centered at $(0, \ldots, 0, -\varepsilon)$ of radius $C \cdot R(\varepsilon)$, where C is a constant depending on the coefficients of $\mathcal{L}$ *but not on* u.

In view of the claim, this polydisk covers the origin for small ε and so the conclusion of the theorem follows.

□

PROOF OF THE CLAIM. According to our normalization, the boundary of Ω near the origin is given by

$$Re\, z_n = \phi\,(z', Im\, z_n)$$

where $\phi \in C^1$, $\phi(0) = 0$, and $d\phi(0) = 0$. Hence, $\phi\,(z', Im\, z_n) = o(|z|)$ near the origin, and the claim follows. □

REMARK 4.1. In Zerner's theorem the hypothesis of Ω being C^1 can be relaxed somewhat. It suffices to require that $\partial\Omega$ has a tangent everywhere and when that tangent plane is moved parallel to itself inside Ω by the distance ε in the normal direction the intersection with Ω contains a polydisk of radius $R(\varepsilon)$ so that

$$\lim_{\varepsilon \to 0} \frac{\varepsilon}{R(\varepsilon)} = 0.$$

REMARK 4.2. Zerner's theorem is a $\mathbb{C}^n$*-theorem only.* Indeed, consider a solution to Laplace's equation in the unit disk in $\mathbb{C}$. The surface

$$\varphi(x, y) := Re\left(x^2 + y^2 - 1\right) = 0$$

in $\mathbb{C}^2$ does not have any characteristic points with respect to the Laplacian, since

$$Re\left\{\left(\frac{\partial\varphi}{\partial x}\right)^2 + \left(\frac{\partial\varphi}{\partial y}\right)^2\right\} = Re\left(4\left(x^2 + y^2\right)\right) = 4.$$

Nevertheless, there are harmonic functions in the unit disk that do not extend across any point on the unit circle.

EXERCISE 4.1. Construct such a function, i.e., find a harmonic function in the unit disk that does not extend past any point on the unit circle.
Hint: Consider the solution u to Dirichlet's problem in the unit disk with a strategic choice of continuous boundary data (where the nature of the data prevents u from extending to a neighborhood of any boundary point).

4. A corollary: The Delassus-Le Roux theorem

COROLLARY 4.1. *If u is holomorphic in a domain Ω in $\mathbb{C}^n$ except for points on a holomorphic hypersurface $\Gamma = \{z : \varphi(z) = 0\}$, and satisfies a PDE $\mathcal{L}u = f$ in $\Omega \setminus \Gamma$ with holomorphic coefficients in Ω then u extends holomorphically to Ω provided Γ is non-characteristic with respect to $\mathcal{L}$. Thus, Γ can only be a non-removable singularity set for u if Γ is (everywhere) characteristic with respect to $\mathcal{L}$.*

PROOF. Imbed Γ into a real hypersurface S given by the equation $Re\,\varphi(z) = 0$. Near Γ S is nowhere Zerner characteristic with respect to $\mathcal{L}$. Indeed, the characteristicity condition is in this case

$$\sum_{|\alpha|=m} a_\alpha(z)\,(\nabla_z\varphi)^\alpha|_\Gamma = 0$$

because $\partial_z \bar{\varphi} = 0$. So all points on $\Gamma \subset S$ satisfy the hypothesis of Zerner's theorem and the corollary follows. □

EXERCISE 4.2. Prove that if the rational function f/g satisfies the Laplace equation, with f and g relatively prime polynomials, then

$$\sum_{j=1}^{n} \left(\frac{\partial g}{\partial z_j} \right)^2 = 0$$

on the variety $\{g = 0\}$.

EXERCISE 4.3. Consider the unbounded domain

$$\Omega := \{z \in \mathbb{C}^2 : \exp\{Re z_1\} + \exp\{Re z_2\} < 1\}.$$

(a) For the operator $\mathcal{L} := \frac{\partial^2}{\partial z_1^2} - \frac{\partial^2}{\partial z_2^2}$, show that any solution u of $\mathcal{L}u = 0$ holomorphic in Ω extends past each point of $\partial\Omega$ except where $Re z_1 = Re z_2$.

(b) For $\mathcal{L} := \frac{\partial^2}{\partial z_1 \partial z_2}$, show that a solution u of $\mathcal{L}u = 0$ holomorphic in Ω extends past *every* point of $\partial\Omega$.

Notes

Zerner's theorem is proved in [**203**]. In the proof presented here we have followed the exposition in Hörmander's book [**86**, Theorem 9.4.7]. The theorem of Delassus and Le Roux can be found, e.g., in Hadamard's book [**72**].

Zerner's theorem has been since extended to some characteristic points as well, cf., e.g., [**42**], [**89**], and references cited therein. Further generalizations can be found in [**7**]. Also, the reader might confer with Ch. 10 where applications of some of these new versions of Zerner's theorem at characteristic points are given.

For a modern coaching of Zerner's theorem in the setting of "microlocal analysis", see, e.g., [**98**]. See also the survey, [**170**], that contains lots of further references.

There have been many attempts to extend Zerner's theorem to nonlinear equations, although one cannot get an equally elegant and simple statement in that context. For references in this directions, see [**133**].

CHAPTER 5

The Method of Globalizing Families

1. Globalizing families

Zerner's Theorem gives a sufficient condition for *locally* extending the solution of a holomorphic PDE.

Suppose that we want to globally continue a solution u of the equation $\mathcal{L}u = f$ holomorphic in a domain $\Omega \subset \mathbb{C}^n$ to a larger domain $\tilde{\Omega}$ containing Ω. Of course, we shall always assume that all the coefficients are holomorphic in $\tilde{\Omega}$.

A simple idea is then to construct a family of *bounded* domains Ω_t, $t \in [0, 1]$ (we shall call it a *globalizing family*-GF) satisfying the following properties

(i) $\Omega_t \subset \Omega_s$ if $t \leq s$;
(ii) $\Omega_1 = \tilde{\Omega}$;
(iii) $\Omega_t \subset \Omega$ for t sufficiently small;
(iv) All boundary points $\partial\Omega_t \setminus \Omega$ for all $t \in [0, 1)$ satisfy the hypothesis of Zerner's theorem (with respect to the operator $\mathcal{L}$);
(v) (*Continuity property*) $\bigcup\limits_{s<t} \Omega_s = \Omega_t$, $\bigcap\limits_{s>t} \Omega_s = \overline{\Omega}_t$.

2. The globalizing principle

The following property associated with globalizing families is known as the *globalizing principle*:

If for given $\mathcal{L}$, Ω, $\tilde{\Omega}$ *we can find a globalizing family satisfying* (i)–(v), *then any solution* u *of* $\mathcal{L}u = f$ *holomorphic in* Ω *extends to* $\tilde{\Omega}$.

Indeed let $E = \{t \in [0, 1] : u \text{ is holomorphic in } \Omega_t\}$. Then

a) E is nonempty because of (iii).
b) E is closed (follows at once from (i) and (v)).
c) E is open in view of (iv), Zerner's theorem, compactness of $\partial\Omega_t$, and (v).

Hence $E = [0, 1]$ and so u extends to $\tilde{\Omega}$ by (ii).

3. Applications

The following simple examples hopefully illustrate well the power of the globalizing method.

COROLLARY 5.1. *Let* u *be holomorphic in a* $\mathbb{C}^n$*-neighborhood of the isotropic cone*

$$\Gamma_0 := \left\{ z \in \mathbb{C}^n : \sum_{j=1}^{n} z_j^2 = 0 \right\}$$

and satisfy the Laplace equation there. Then u *extends as an entire function to all of* $\mathbb{C}^n$.

PROOF. The corollary follows from the Globalizing Principle if we choose the following globalizing family:

$$\Omega_t = \left\{ z : \sum_{j=1}^{n} |z_j|^2 \le t^2 \right\}, \quad t \in [0, \infty).$$

□

EXERCISE 5.1. Fill in the details in the above proof, noting that the only characteristic points on $\partial\Omega_t$ are contained in Γ_0. Note that although the setup is slightly modified from the above statement (here the union of domains Ω_t results in an unbounded domain, and the interval is $[0, \infty)$ instead of $[0, 1]$), but the same argument (based on showing that E is both open and closed) applies.

COROLLARY 5.2. *The solution of the Cauchy-Goursat problem*

$$\begin{cases} \partial^\beta w &= \sum_{|\alpha|<m} a_\alpha(z)\partial^\alpha w + f, |\beta| = m, \\ w &= \cdots = \partial^{\beta_1 - 1} w = 0 \ on \ \{z_1 = 0\}, \\ \cdot & \cdots \quad \cdots \\ w &= \cdots = \partial^{\beta_n - 1} w = 0 \ on \ \{z_n = 0\}, \end{cases}$$

is an entire function provided all the coefficients a_α and the function f are entire.

PROOF. We are going to prove this with a particular choice of $\beta = (0, 0, \ldots, m)$, leaving the general case as an exercise.

Let us define the globalizing family to be the family of concentric balls

$$\Omega_t = \{z : \|z\| < t, 0 \le t < \infty\}.$$

It is easily seen that a point on $\partial\Omega_t$ is characteristic if and only if $\overline{z}_n^m = 0$, and so characteristic points are all located in the initial hyperplane.

Near those points the solution is holomorphic by the Cauchy-Kovalevskaya theorem. For all other points on $\partial\Omega_t$, w extends across by Zerner's theorem. Thus, again letting $E = \{t \in [0, \infty) : w \text{ is holomorphic in } \Omega_t\}$, it follows as before that $E = [0, \infty)$ and so w is an entire function. □

The following particular case of Corollary 5.2 is of the utmost importance (cf. Ch 3, Sec. 5).

COROLLARY 5.3. *The Riemann function $R(z, w; \xi, \eta)$ for the operator*

$$\mathcal{L} := \frac{\partial^2}{\partial z \partial w} + a(z, w)\frac{\partial}{\partial z} + b(z, w)\frac{\partial}{\partial w} + c(z, w)$$

is entire for any point (ξ, η) provided that all the coefficients are entire functions.

EXAMPLE 5.1. Let $\mathcal{L} = \frac{\partial^2}{\partial z \partial w} + \lambda^2$ be the Helmholtz operator. Then, for the Riemann function we have (cf. [**72**])

$$R(z, w; \xi, \eta) = J_0\left(\lambda\sqrt{(z - \xi)(w - \eta)}\right),$$

where J_0 is the Bessel function of order zero [**200**] given by

$$J_0(x) = \sum_{n=0}^{\infty} \frac{(-1)^n}{(n!)^2} \left(\frac{x}{2}\right)^{2n}.$$

COROLLARY 5.4 (The generalized heat equation). *If u is a holomorphic solution of the equation*

$$\frac{\partial^2 u}{\partial z^2} + a(z,w)\,\frac{\partial u}{\partial w} + b(z,w)\,\frac{\partial u}{\partial z} + c(z,w) = f(z,w)$$

in a bidisk $\mathcal{D}(0,R)$ and a,b,c,f are entire functions then u extends as a holomorphic function to the "cylinder"

$$\{(z,w) : z \in \mathbb{C}, |w| < R\}.$$

PROOF. Consider the family of ellipsoids

$$\Omega_t = \left\{ (z,w) : \frac{|z|^2}{t^2} + \frac{|w|^2}{(R')^2} \leq 1 \right\},$$

where $R' < R$ is fixed.

To check that $\{\Omega_t\}_{t \geq 0}$ is a globalizing family we only have to verify the condition (iv) in the definition of (GF). The defining function for the boundary $\partial\Omega_t$ is

$$\varphi(z,w) = \frac{z\bar{z}}{t^2} + \frac{w\bar{w}}{(R')^2} - 1 = 0.$$

So the characteristic points are precisely those for which $z = 0$. But these points are in $D(0,R)$ where u is holomorphic. The rest of the argument follows *mutatis mutandis.* □

REMARK 5.1. In particular, the conclusion of Corollary 5.4 entails that the "data" $u(z,0)$ must be entire. This provides another perspective on Kovalevskaya's example discussed in Section 2 of Chapter 3. Namely, the particular choice of f is not important. It is merely the fact that f is not entire that prevents the solution from having a positive radius of convergence.

Moreover, a deeper analysis in the case when, e.g., a,b,c and f are all polynomials yields a much more accurate result: for any fixed $w : |w| < R$, $u(z,w)$ is an entire function of z of order at most 2. We shall leave the details as an exercise for an ambitious reader.

EXERCISE 5.2. Let $\mathcal{L} := \frac{\partial^2}{\partial z_1 \partial z_2}$ and consider the unbounded domain

$$\Omega := \{z \in \mathbb{C}^2 : \exp\{Rez_1\} + \exp\{Rez_2\} < 1\}.$$

Suppose a solution u of $\mathcal{L}u = 0$ is holomorphic in Ω, and find a globalizing family of domains that extends u to all of $\mathbb{C}^n$.

Notes

The method of globalizing families in various editions based on a continuous deformation of the boundary of a domain of regularity has appeared in works of many authors: F. John [**93**] and L. Hörmander [**86**] used it for proving various uniqueness theorems in $\mathbb{R}^n$; J. Bony and P. Schapira [**22**] applied this idea to extend the domain of regularity of the solution of a PDE in $\mathbb{C}^n$ (also, cf. [**86**, Theorem 9.4.8]. The method has been developed further and used extensively by G. Johnsson [**94**]. Here, we followed the outline given in [**103**]. Corollary 5.1 is due to G. Johnsson (private communication). Proofs of Corollaries 5.2 and 5.3 (but not the facts) are due to the first author. For a "classical" proof of Corollary 5.3 see, e.g., Vekua's book [**196**]. Corollary 5.4 for the heat equation was observed by

Kovalevskaya in her Habilitationsschrift [**119**]. An extended version and the proof given here follow [**114**].

CHAPTER 6

Holmgren's Uniqueness Theorem

The Cauchy-Kovalevskaya theorem provides existence and uniqueness within the class of holomorphic functions. This raises the question of whether uniqueness still holds within a more general class of, say, smooth solutions. Holmgren's theorem, presented in Section 2 of this chapter, addresses this question.

1. A uniqueness result for harmonic functions

Let us start out by recalling a simple uniqueness theorem for harmonic functions (see [**99**, Ch. X, Sec. 5]).

THEOREM 6.1. *Let Ω be a domain in $\mathbb{R}^n$ with C^1-smooth boundary, and Γ be a C^1-smooth hypersurface inside Ω that divides it into two parts: Ω^+ and Ω^-. Let $u \in C^2(\Omega^+ \cup \Omega^-) \cap C^1(\overline{\Omega})$. Suppose $\Delta u = 0$ in $\Omega^+ \cup \Omega^- = \Omega \setminus \Gamma$. Then u extends as a harmonic function to all of Ω.*

COROLLARY 6.1. *If u is harmonic in Ω^+, $u \in C^1\left(\overline{\Omega^+}\right)$ and each of $u\big|_\Gamma$ and $\nabla u\big|_\Gamma$ vanish, then u vanishes identically.*

PROOF OF THE COROLLARY. Set $u\big|_{\Omega^-} \equiv 0$. By Theorem 6.1, u is harmonic in Ω and thus $u \equiv 0$. □

REMARK 6.1. This is an analogue for harmonic functions of a well-known result on removable singularities of analytic functions based on Morera's theorem [**6**, p. 122]: let $\Omega, \Gamma \subset \mathbb{R}^2$ be as in Theorem 6.1 and let f be analytic in $\Omega^+ \cup \Omega^-$ and continuous in $\overline{\Omega}$. Then f extends as an analytic function to all of Ω.

PROOF OF THEOREM 6.1. For the sake of simplicity take $n = 3$.
Fix $x \in \Omega^+$, we have, by Green's formula,

$$u(x) = \frac{1}{4\pi}\int_{\partial\Omega^+}\left[u(y)\frac{\partial}{\partial n_y}\left(|x-y|^{-1}\right) - \frac{\partial u}{\partial n_y}\left(|x-y|^{-1}\right)\right]dS_y, \tag{6.1}$$

where n is the outer normal and dS_y is Lebesgue measure on $\partial\Omega^+$.
Also, by Green's formula,

$$0 = \frac{1}{4\pi}\int_{\partial\Omega^-}\left[u(y)\frac{\partial}{\partial n_y}\left(|x-y|^{-1}\right) - \frac{\partial u}{\partial n_y}\left(|x-y|^{-1}\right)\right]dS_y.$$

Hence, in (6.1) the part of the integral over Γ can be replaced by that over $\partial\Omega^- \setminus \Gamma$. Substituting this into (6.1) gives the extension of u into all of Ω. □

EXERCISE 6.1. Extend the above proof for all $n > 1$.

2. Holmgren's uniqueness theorem

Holmgren's uniqueness theorem is a far reaching extension of Corollary 6.1 to solutions of arbitrary equations with real-analytic coefficients. Here, for the sake of simplicity, we shall present it in a slightly reduced form.

THEOREM 6.2. *Let Ω be a domain in $\mathbb{R}^n$ with C^1 boundary, $\Omega \subset \{x_n > 0\}$, and $\{x' : \|x'\| < R\} \subset \partial\Omega \cap \{x_n = 0\}$. Let $w \in C^m(\Omega) \cap C^{m-1}\left(\overline{\Omega}\right)$ be a solution of*

$$\partial_n^m w = \sum_{\substack{|\alpha| \le m \\ \alpha \ne (0,\ldots,0,m)}} a_\alpha(x) \partial^\alpha w$$

in Ω and $w \mid_{\{x_n=0\}} = \cdots = \partial_n^{m-1} w \mid_{\{x_n=0\}} = 0$. Then, if all a_α's are real-analytic in $\overline{\Omega}$, there exists $\delta > 0$ such that $w \equiv 0$ in $\{x : \|x\| \le \delta\} \cap \Omega$.

The idea of the proof comes from ODEs.

Indeed, let us discuss the following example. Consider a solution u of the ODE $u'' + pu' + qu = 0$ on $[0,1]$, with the coefficients $p, q \in C^1[0,1]$ satisfying $u(0) = u'(0) = 0$. Then, we claim that $u \equiv 0$ on $[0,1]$.

Take $0 < a < 1$. We want to show $u \mid_{[0,a]} \equiv 0$. Take any $v \in C^1[0,a]$ such that $v(a) = v'(a) = 0$. Integration by parts yields:

$$\int_0^a u'' v \, dx = u'v|_0^a - \int_0^a u'v' dx = \int_0^a uv'' dx,$$

$$\int_0^a pu'v \, dx = -\int_0^a u(pv)' \, dx.$$

So denoting by L the ordinary differential operator $L : \dfrac{d^2}{dx^2} + p\dfrac{d}{dx} + q$, we have

$$\int_0^a (Lu)v \, dx = 0 = \int_0^a u\left(v'' - (pv)' + q\right) dx = \int_0^a uL^\sharp v \, dx,$$

where $L^\sharp := \dfrac{d^2}{dx^2} + p\dfrac{d}{dx} + (q - p')$ denotes the adjoint operator.

By the Main Existence Theorem for ODEs [**90**, Sec. 3.3] we can solve the initial value problem

$$\begin{cases} L^\sharp v = f; \\ v(a) = v'(a) = 0 \end{cases}$$

on $[0,1]$ for any reasonable function f, say a polynomial.

Therefore $\int_0^a u\, f\, dx = 0$ for all polynomials f and hence $u \equiv 0$.

PROOF OF THE THEOREM. For the sake of simplicity, we take up the case $m = n = 2$. We have

$$\frac{\partial^2 w}{\partial y^2} = -a(x,y)\frac{\partial^2 w}{\partial x^2} - b(x,y)\frac{\partial^2 w}{\partial x \partial y} - c(x,y)\frac{\partial w}{\partial x} - d(x,y)\frac{\partial w}{\partial y} - e(x,y)w,$$

where a, b, c, d, e are analytic in Ω.

Consider $C_\varepsilon := \left\{(x,y) : y + x^2 = \varepsilon\right\}$ and let $\Omega_\varepsilon \subset \Omega$ be a domain bounded by C_ε and $\{y = 0\}$. Take $v \in C^2(\Omega_\varepsilon) \cap C^1\left(\overline{\Omega_\varepsilon}\right)$ such that v, ∇v vanish on C_ε.

CLAIM. For any $u \in C^2(\Omega_\varepsilon) \cap C^1(\overline{\Omega_\varepsilon})$ such that u, ∇u vanish on $\partial\Omega_\varepsilon \cap \{y=0\}$ we have

$$\iint_{\Omega_\varepsilon} (\mathcal{L}u)v\,dx\,dy = \iint_{\Omega_\varepsilon} u\left(\mathcal{L}^\sharp v\right) dx\,dy = 0,$$

where

$$\mathcal{L} = \frac{\partial^2}{\partial y^2} + a(x,y)\frac{\partial^2}{\partial x^2} + b(x,y)\frac{\partial^2}{\partial x\partial y} + c(x,y)\frac{\partial}{\partial x} + d(x,y)\frac{\partial}{\partial y} + e(x,y)$$

and $\mathcal{L}^\sharp$ is the adjoint operator

$$\mathcal{L}^\sharp w = \frac{\partial^2 w}{\partial y^2} + \frac{\partial^2(aw)}{\partial x^2} + \frac{\partial^2(bw)}{\partial x\partial y} - \frac{\partial(cw)}{\partial x} - \frac{\partial(dw)}{\partial y} + ew.$$

PROOF OF CLAIM. Note that all the products $uv, u_z v, \ldots, u_y v_y$ vanish on $\partial\Omega_\varepsilon$.

The proof of the claim is now obtained by a straightforward computation involving integration by parts. For the reader's convenience we shall perform it for two of the terms involved.

Integrating by parts twice, we obtain

$$\begin{aligned}
\iint_{\Omega_\varepsilon} bv\frac{\partial^2 u}{\partial x\partial y}\,dx\,dy &= \int_{-\sqrt{\varepsilon}}^{\sqrt{\varepsilon}}\left\{\int_0^{\varepsilon-x^2} bv\frac{\partial^2 u}{\partial x\partial y}\,dy\right\}dx\\
&= -\int_{-\sqrt{\varepsilon}}^{\sqrt{\varepsilon}}\left\{\int_0^{\varepsilon-x^2} \frac{\partial(bv)}{\partial y}\frac{\partial u}{\partial x}\,dy\right\}dx\\
&= -\int_0^{\varepsilon}\left\{\int_{-\sqrt{\varepsilon-y}}^{\sqrt{\varepsilon-y}} \frac{\partial(bv)}{\partial y}\frac{\partial u}{\partial x}\,dx\right\}dy\\
&= \int_0^{\varepsilon}\left\{\int_{-\sqrt{\varepsilon-y}}^{\sqrt{\varepsilon-y}} \frac{\partial^2(bv)}{\partial x\partial y}u\,dx\right\}dy\\
&= \iint_{\Omega_\varepsilon} u\frac{\partial^2(bv)}{\partial x\partial y}\,dx\,dy.
\end{aligned}$$

Similarly,

$$\begin{aligned}
\iint_{\Omega_\varepsilon} cv\frac{\partial u}{\partial x}\,dx\,dy &= \int_0^{\varepsilon}\left\{\int_{-\sqrt{\varepsilon-y}}^{\sqrt{\varepsilon-y}} cv\frac{\partial u}{\partial x}\,dx\right\}dy\\
&= -\int_0^{\varepsilon}\left\{\int_{-\sqrt{\varepsilon-y}}^{\sqrt{\varepsilon-y}} \frac{\partial(cv)}{\partial x}u\,dx\right\}dy\\
&= -\iint_{\Omega_\varepsilon} u\frac{\partial(cv)}{\partial x}\,dx\,dy.
\end{aligned}$$

□

Provided that C_ε is not characteristic with respect to $\mathcal{L}^\sharp$ for small ε, we can solve the Cauchy problem

$$\begin{cases} \mathcal{L}^\sharp v = p(x,y),\\ v = 0, \nabla v = 0 \text{ on } C_\varepsilon \end{cases}$$

with $p(x,y)$ being an arbitrary polynomial.

By the C-K theorem (cf. Chapter 3), the solution v exists in a domain D of size $const\sqrt{\varepsilon}$ which therefore covers the origin. Now the claim implies that for all sufficiently small ε, $\iint_{\Omega_\varepsilon} u \cdot p(x,y)\,dx\,dy = 0$ for any polynomial p and therefore $u \equiv 0$ in Ω_ε.

To check that C_ε is non-characteristic for $\mathcal{L}^\sharp$ we note that C_ε is given by the equation $C_\varepsilon : \varphi(x,y) := x^2 + y - \varepsilon = 0$ and the characteristic (with respect to $\mathcal{L}^\sharp$) form for the function φ can be written as

$$Char\left(\mathcal{L}^\sharp, C_\varepsilon\right) = 1 + \left(\text{terms containing } \frac{\partial\varphi}{\partial x}\right).$$

Since $\dfrac{\partial\varphi}{\partial x} = 0$ at $(0,0)$, C_ε is non-characteristic for small ε. □

The following exercise (for an ambitious reader) establishes Holmgren's theorem in full generality.

EXERCISE 6.2. Extend Holmgren's theorem to more general operators

$$\mathcal{L} = \sum_{|\alpha|\le m} a_\alpha(x)\partial^\alpha$$

with holomorphic coefficients and the "data" surface $\{x_n = 0\}$ replaced by an arbitrary C^1-smooth surface Γ that is non-characteristic with respect to $\mathcal{L}$.

The hypothesis that the initial hypersurface is non-characteristic is crucial for Holmgren's theorem. The following exercise shows this by constructing a whole family of smooth solutions (depending on a parameter a) to the heat equation that vanish for $t \le 0$ but not for $t > 0$.

EXERCISE 6.3 (A. Tychonoff's example [**194**]). The steps in this exercise follow [**93**, Ch. VII, p. 211]. (a) Write the solution of the one-dimensional heat equation $u_t = u_{xx}$ with data on the t-axis

$$u(0,t) = g(t),$$
$$u_x(0,t) = 0,$$

as a power series:

$$u(x,t) = \sum_{j=0}^{\infty} g_j(t)x^j.$$

Show that to satisfy the equation one must have

$$g_0 = g,$$
$$g_1 = 0,$$
$$g_j' = (j+2)(j+1)g_{j+2},$$

hence leading to the formal solution

$$u(x,t) = \sum_{k=0}^{\infty} \frac{g^{(k)}(t)}{(2k)!}x^{2k}. \tag{6.2}$$

(b) Choose, for some $a > 1$, the function $g(t)$ defined by

$$g(t) = \begin{cases} \exp\left(-t^{-a}\right), & t > 0, \\ 0, & t \le 0. \end{cases}$$

Show that there exists $\theta = \theta(a) > 0$ such that for all $t > 0$

$$|g^{(k)}(t)| \le \frac{k!}{(\theta t)^k} e^{-\frac{1}{2}t^{-a}}. \tag{6.3}$$

Hint: Use Cauchy estimates for $g^{(k)}$. More precisely, fixing $t_0 > 0$ write the Cauchy integral formula for the kth derivative at t_0, using as a path of integration in the complex t-plane a circle C centered at t_0 of radius $\theta(a) \cdot t_0$, with $\theta > 0$ so small that $Re(t^{-a}) > \frac{1}{2}t_0^{-a}$ for all $t \in C$.

(b′) Show that there exist M, r only depending on a such that for all k and for all $t > 0$,

$$|g^{(k)}(t)| \leq M \cdot (k!)^{1+\frac{1}{a}} r^{-k}. \tag{6.4}$$

(c) Show that for $t > 0$ and any complex x

$$\sum_{k=0}^{\infty} \left| \frac{g^{(k)}(t)}{(2k)!} x^{2k} \right| \leq \sum_{k=0}^{\infty} \frac{|x|^{2k}}{k!(\theta t)^k} e^{-t^{-a}/2} = \exp\left\{ \frac{1}{t} \left(\frac{|x|^2}{\theta} - \frac{t^{1-a}}{2} \right) \right\}, \tag{6.5}$$

and also trivially for $t < 0$. **Hint:** $\dfrac{k!}{(2k)!} < \dfrac{1}{k!}$.

Conclude that the series (6.2) converges for all $t \in \mathbb{R}$ and complex x. Moreover, the estimate (6.5) implies that

$$\lim_{t \to 0^+} u(x, t) = 0,$$

uniformly as x ranges over any compact set.

(d) Set

$$U(x, t) = \begin{cases} \exp\left\{ \frac{1}{t} \left(\frac{|x|^2}{\theta} - \frac{t^{1-a}}{2} \right) \right\}, & t > 0, \\ 0, & t \leq 0. \end{cases}$$

The series (6.2) is majorized by $U(x, t)$, and as x ranges over any compact set in $\mathbb{C}$, $U(x, t)$ is bounded (uniformly for all $t \in \mathbb{R}$). Conclude that (6.2) converges uniformly in x, t for bounded x and all $t \in \mathbb{R}$.

(e) Note that the above conclusion also holds for the series obtained by term by term x-differentiation of (6.2). Thus, the series

$$\sum_{k=2}^{\infty} \frac{g^{(k)}(t)}{(2k-2)!} x^{2k-2} = \sum_{k=0}^{\infty} \frac{g^{(k+1)}(t)}{(2k)!} x^{2k} \tag{6.6}$$

converges uniformly. Since the latter series is obtained from (6.2) by formal differentiation with respect to t, $u_t = u_{xx}$. More generally,

$$\frac{\partial^k}{\partial t^k} u = \frac{\partial^{2k}}{\partial x^{2k}} u,$$

which implies that $u \in C^\infty(\mathbb{R}^2)$.

Thus, $u(x, t)$ is an entire function of x (of what order?) for any real t but, of course, is not analytic in t, since $u(0, t) \equiv 0$ for $t \leq 0$ but not for $t > 0$.

Notes

The Holmgren uniqueness theorem (Theorem 6.2) was proved by Holmgren [**84**] in a special case and by F. John [**92**] in full generality (Exercise 6.2). Also, see [**86**], [**93**]. The discussion of a rather special case presented here follows Petrovsky [**151**]. Theorem 6.1 is from Kellogg's classic [**99**, Ch. X, Sec. 5]. Folklore attributes the result to Kovalevskaya.

The example presented in Exercise 6.3 is due to A. Tychonoff [**194**]. Here we have followed the presentation from F. John's book [**93**, Ch. VII p. 211]. See also

J. Hadamard's lectures [**72**]. Functions satisfying (6.4) are said to belong to Gevrey class $1 + 1/a$. Real analytic functions are of class 1.

CHAPTER 7

The Continuity Method of F. John

1. A global uniqueness result

Sometimes one can use the method of globalizing families to extend Holmgren's uniqueness theorem, which is local in nature. The following example serves as a good illustration of the methods one may use.

THEOREM 7.1. *Let a C^2-smooth function u satisfy the following equation in $\mathbb{R}^{n+1}$:*

$$\frac{\partial^2 u}{\partial t^2} = \sum_{j=1}^{n} \frac{\partial^2 u}{\partial x_j^2} + \sum_{j=1}^{n} a_j(x,t)\,\frac{\partial u}{\partial x_j} + a_{n+1}(x,t)\,\frac{\partial u}{\partial t} + a_0(x,t)u$$

in the double cone $C = \{(x,t) : \|x\| + |t| < R\}$ and have zero Cauchy data on $\{(x,0) : \|x\| < R\}$ (the base of the cone). Then $u \equiv 0$ provided that all the coefficients $a_j = a_j(x,t)$ are real analytic in a neighborhood of C.

For the sake of simplicity, we shall prove the theorem assuming in addition that $n = 1$ and $R = 1$. The following lemma is elementary and the proof is left as an exercise.

LEMMA 7.1. *Fix a point (x_0, t_0) in C with $t_0 > 0$. There exists a function f on $[-1,1]$ satisfying the following properties:*

(i) $f(-1) = f(1) = 0$;
(ii) $f \in C^1[-1,1]$;
(iii) $f(x_0) > t_0$;
(iv) $|f'(x)| < 1$ on $[-1,1]$.

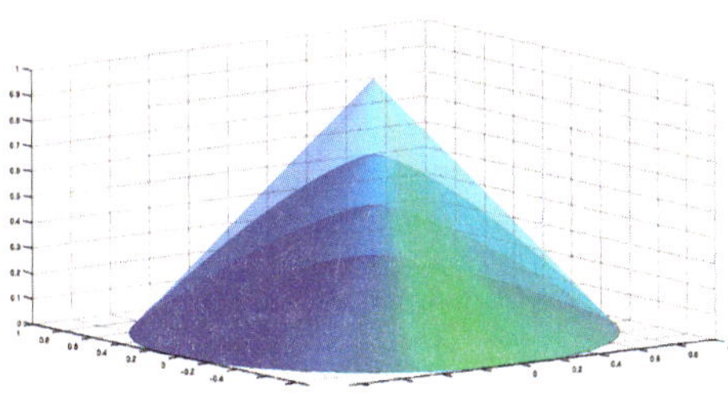

FIGURE 1. As s ranges from 0 to 1, the family of nowhere-characteristic surfaces $t = sf(x)$ sweeps out the upper-half of C.

The usefulness of Lemma 7.1 is immediately explained in Fig. 1.

PROOF OF THEOREM 7.1. For $s \in [0,1]$ define

$$\mathcal{D}_s = \{(x,t) : t - sf(x) < 0\},$$

where f is as in Lemma 7.1. Clearly $(x_0, t_0) \in \mathcal{D}_1$.

Let $E = \{s \in [0,1] : u \equiv 0 \text{ in } \mathcal{D}_s\}$. We shall show that $E = [0,1]$. Indeed,

(i) E is nonempty in view of Holmgren's theorem;

(ii) E is obviously closed. For if $s_n \to s_0$, $\{s_n\}_{n=1}^{\infty} \subset E$ then

$$\mathcal{D}_{s_0} \subset \bigcup_{n=1}^{\infty} \mathcal{D}_{s_n}$$

and so $u \equiv 0$ in $\mathcal{D}_{s_0}$, implying that $s_0 \in E$.

(iii) E is open. It is again a direct consequence of Holmgren's uniqueness theorem provided that we check that the part of the boundary of $\mathcal{D}_s$ given by the curve $t = sf(x)$ is non-characteristic. But this is obvious since setting $\varphi(x,t) := t - sf(x)$ we have

$$\left(\frac{\partial \varphi}{\partial x}\right)^2 - \left(\frac{\partial \varphi}{\partial t}\right)^2 = s^2 \left|f'(x)\right|^2 - 1 < 0$$

because of property (iv) in Lemma 7.1 satisfied by the function f.

So we have $E = [0,1]$, and $u \equiv 0$ in the upper-half of the double cone. The argument for the lower half of C is the same. □

2. Exercises

EXERCISE 7.1. Prove Lemma 7.1.

The following exercises serve as a reminder that Holmgren's uniqueness theorem is local and in an essential way rests on assumptions of real analyticity for both coefficients and the initial data.

EXERCISE 7.2. Consider the Cauchy problem for the one-dimensional wave equation

$$\begin{cases} u_{tt} - u_{xx} = 0, \\ u(x,0) = f(x), \\ u_t(x,0) = g(x). \end{cases} \tag{7.1}$$

Show that for $f \in C^2$ and $g \in C^1$ the solution to (7.1) can be represented by d'Alembert's formula

$$u(x,t) = \frac{1}{2}\left[f(x-t) + f(x+t)\right] + \frac{1}{2}\int_{x-t}^{x+t} g(y)dy. \tag{7.2}$$

Hint: Rewrite the wave equation in terms of the variables $\xi = x + t$, $\eta = x - t$, and show that the solution is of the form $F(\xi) + G(\eta)$, or changing back to x and t:

$$u(x,t) = F(x+t) + G(x-t). \tag{7.3}$$

Use (7.3) and the initial conditions in order to derive (7.2).

The representation (7.2) implies that the value of the solution $u(x,t)$ only depends on the values of f and g in the interval $[x-t, x+t]$. This interval is referred to as the *domain of dependence* for the solution at the point (x,t).

EXERCISE 7.3. Taking the case $R = 1$, use the results of the previous exercise to describe a solution u of the wave equation (7.1) that is C^2-smooth throughout $\mathbb{R}^2$, vanishes identically in the cone $C := \{|x| + |t| < 1\}$, but is nonzero arbitrarily close to C.

EXERCISE 7.4. (Propagation of singularities - "shock waves".) Consider the initial value problem (7.1) with $g(x) = 0$ and

$$f(x) = \begin{cases} 1 - |x|, & |x| \leq 1, \\ 0, & |x| > 1. \end{cases}$$

Even though it fails to be smooth, interpret (7.2) as a so-called weak solution.

(a) Describe all lines of the form $t = mx + b$ on which singularities propagate (in this case, a singularity refers to a point where $u(x, t)$ fails to be differentiable).

(b) Label all regions in the upper halfplane over which $u(x, t) = 0$.

(c) For large fixed $t > 0$ sketch the graph of $u(x, t)$ (that is, the graph of $u(x, t)$ with respect to x, treating t as a parameter), and describe how the graph changes with respect to t.

Notes

The method of proof of Theorem 7.1 is due to F. John [**93**], also, cf. [**85**]. Similar ideas were used in Chapter 5 to extend the domain of *existence* of solutions of analytic PDE's. This theme will be further developed in the following chapters.

CHAPTER 8

The Bony-Schapira Theorem

Suppose w is a solution to $\mathcal{L}w = f$ in $\Omega_1 \subset \Omega_2$. Zerner's theorem provided a means for locally extending the domain of w, and in Chapter 5 we saw that it is possible to extend w to all of Ω_2 provided there is a globalizing family of intermediate domains. This raises the question of whether there are any simple geometric conditions on Ω_1 and Ω_2 which guarantee the existence of a globalizing family. A convenient sufficient condition is provided by the following theorem.

THEOREM 8.1. *Let $\Omega_1 \subset \Omega_2$ be convex, bounded domains in $\mathbb{C}^n$. Let*

$$\mathcal{L} = \sum a_\alpha(z)\partial^\alpha$$

be a differential operator whose coefficients a_α are holomorphic in $\bar{\Omega}_2$. Every w holomorphic in $\bar{\Omega}_1$ and satisfying $\mathcal{L}w = f$, with f holomorphic in Ω_2, extends holomorphically to Ω_2 provided the following condition holds: for each $z \in \Omega_2$ and each real hyperplane H passing through z and characteristic with respect to $\mathcal{L}$ at z, $H \cap \Omega_1 \neq \emptyset$.

By *characteristic* we mean *Zerner characteristic* as defined in Chapter 4.

Recall that a set is said to be *convex* if for every pair of points $z^{(1)}, z^{(2)}$ in the set, the line segment $(1-t)z^{(1)} + tz^{(2)}$ is also contained in the set. The *convex hull* of a set is the intersection of all convex sets containing it, i.e., the smallest *convex* set containing the given set.

1. Applications of the Bony-Schapira theorem

Before proving the Bony-Schapira Theorem, it seems appropriate to demonstrate how it can be useful. (More elaborate applications will follow in Chapters 9 and 10.)

COROLLARY 8.1. *Assume that all leading coefficients are constants, i.e., $a_\alpha =$ const, $|\alpha| = m$. If Ω_1, Ω_3 are convex, open sets in $\mathbb{C}^n$ such that $\Omega_1 \cap \Omega_3 \neq \emptyset$ and every characteristic real hyperplane H that meets Ω_3 also meets Ω_1, then every solution w of $\mathcal{L}w = f$ holomorphic in $\bar{\Omega}_1$ extends to $\Omega_2 := co\,(\Omega_1 \cup \Omega_3) :=$ $\{$convex hull of $\Omega_1 \cup \Omega_3\}$, provided that all coefficients are holomorphic in $\bar{\Omega}_2$.*

PROOF OF COROLLARY 8.1. We have Ω_2 contains Ω_1 and is convex so it remains to check the hypothesis of the theorem. Let $z^0 \in \Omega_2$ and $H \ni z^0$ be a real hyperplane so that H is characteristic with respect to $\mathcal{L}$. (Since all the senior coefficients are constants, a hyperplane characteristic at one point maintains this property everywhere.) We have three possibilities:

(i) $H \cap \Omega_3 \neq \emptyset$, and hence by the hypothesis H meets Ω_1;
(ii) If $H \cap \Omega_1 \neq \emptyset$ there is nothing to prove;

(iii) If $H\cap\Omega_1=\emptyset$, $H\cap\Omega_3=\emptyset$ then because $\Omega_1\cup\Omega_3$ is connected, Ω_1,Ω_3 both lie in the same half-space with respect to H. Therefore $\Omega_2=co\,(\Omega_1\cup\Omega_2)$ lies in the same half-space contradictory to our assumption that $\Omega_2\cap H\neq\emptyset$.

□

COROLLARY 8.2. *Let u be harmonic in the ball of radius R in $\mathbb{R}^n$, $n\geq 2$ centered at the origin. Then it extends as a holomorphic function to the ball $B\left(0,\frac{R}{\sqrt{2}}\right)$ in $\mathbb{C}^n$ and the constant $\frac{R}{\sqrt{2}}$ is the best possible, i.e., u need not extend to any ball $B\,(0,R')$ in $\mathbb{C}^n$ with $R'>\frac{R}{\sqrt{2}}$.*

PROOF OF COROLLARY 8.2. Without loss of generality we can assume $R=1$ and u to be harmonic in the closed unit ball $\overline{B_{\mathbb{R}^n}(0,1)}$. Since u is harmonic in $\overline{B_{\mathbb{R}^n}(0,1)}$ it is holomorphic in a closed convex $\mathbb{C}^n$ neighborhood $\bar{\Omega}_1$ of $B_{\mathbb{R}^n}(0,1)$. Let $\Omega_3=B_{\mathbb{C}^n}\left(0,\frac{1}{\sqrt{2}}\right)$. By Corollary 8.1 it suffices to check that for every z^0 with $\left|z^0\right|<\frac{1}{\sqrt{2}}$ and any characteristic plane $H:\sum\limits_1^n a_jx_j+b_jy_j=t$ passing through z^0, we have $H\cap B_{\mathbb{R}^n}(0,1)\neq\emptyset$. The characteristicity of H (with respect to the Laplacian) is equivalent to

$$\sum_1^n a_j^2=\sum_1^n b_j^2 \text{ and } \sum_1^n a_jb_j=0.$$

Hence, without loss of generality, we can rescale the equation of H so that $\|a\|=\|b\|=1$. If $H\cap B_{\mathbb{R}^n}(0,1)=\emptyset$, then for any $x\in\mathbb{R}^n$ with $\langle a,x\rangle=t$ it follows that $\|x\|\geq 1$. Hence $|t|\geq 1$, otherwise we could simply take $x=ta$. Now since H passes through z^0, we have (where $z^0=x^0+iy^0$):

$$\begin{aligned}1\leq |t|&=\left|\langle a,x^0\rangle+\langle b,y^0\rangle\right|\\&\leq\left\|x^0\right\|+\left\|y^0\right\|\\&\leq\sqrt{2\left(\left\|x^0\right\|^2+\left\|y^0\right\|^2\right)}\\&=\sqrt{2}|z^0|,\end{aligned}$$

so that $\left|z^0\right|\geq\frac{1}{\sqrt{2}}$. The proof that the constant $\frac{1}{\sqrt{2}}$ is the best possible is left as an exercise (Exercise 8.3 below). □

REMARK 8.1. The same conclusion holds for any operator

$$\mathcal{L}=\sum_1^n\frac{\partial^2}{\partial z_j^2}+\text{(lower order terms)},$$

provided that the coefficients are holomorphic in $\overline{B_{\mathbb{C}^n}(0,R)}$.

In particular, we obtain the following

COROLLARY 8.3. *If u satisfies the equation*

$$\sum_1^n\frac{\partial^2 u}{\partial z_j^2}+\sum_{|\alpha|\leq 1}a_\alpha(z)\partial^\alpha u=f$$

in a $\mathbb{C}^n$-neighborhood of $\mathbb{R}^n$ and all a_α's and f are entire functions, u extends as an entire function to all of $\mathbb{C}^n$.

2. Proof of the Bony-Schapira theorem

Let us now turn to the proof of the theorem.

PROOF OF THEOREM 8.1. Let $z'' \in \Omega_2 \setminus \bar{\Omega}_1$. Take $z' \in \Omega_1$, consider the segment $[z', z''] := \{z_t := tz' + (1-t)z'' : 0 \leq t \leq 1\} \subset \Omega_2$. Take a convex neighborhood $U \supset \Omega_1$, so that u is holomorphic in $\bar{U}$. Without ambiguity we shall denote $U := \Omega_2 \cap U$ and can assume that $\partial U \cap \Omega_2$ is smooth. Choose $\delta > 0$ so that $B_t := \overline{B_{\mathbb{C}^n}(z_t, \delta)} \subset \Omega_2$, for all $z_t \in [z', z'']$. Let D_t denote the convex hull of $\bar{U}$ and B_t. For small t, $D_t = \bar{U}$. Now the argument proceeds by applying the method of globalizing families (cf. Chapter 5).

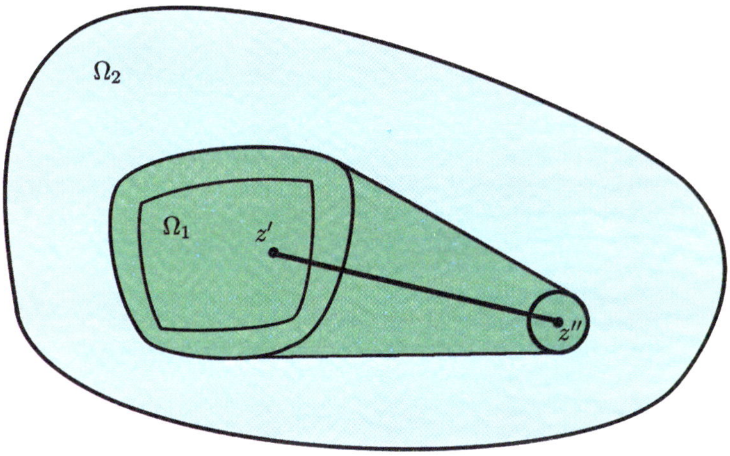

FIGURE 1. Each domain D_t in the GF is constructed as a convex hull of the smooth neighborhood $U \supset \Omega_1$ and a ball B_t of fixed radius with center moving along the segment $[z', z'']$.

We want to show that u extends to all (D_t)'s, $0 \leq t \leq 1$ by showing that the set $E := \{t : 0 \leq t \leq 1 : u \text{ extends holomorphically to } D_t\}$ is nonempty, open, and closed. We notice immediately that the family $\{D_t\}_{0 \leq t < 1}$ satisfies properties (i)–(iii) and (v) of a Globalizing Family, implying that E is both nonempty and closed. Thus, as usual, the only difficulty is to check that the set $\{t : 0 \leq t \leq 1 : u \text{ extends to } D_t\}$ is open.

For any t^0, all points on $\partial D_{t^0} \setminus \bar{U}$ are non-characteristic with respect to $\mathcal{L}$ since a real tangent plane at any such boundary point does not meet $U \cap \Omega_2$, thus it does not meet Ω_1 either, and hence cannot be characteristic by the hypothesis. Hence, by Zerner's theorem u extends across all points on $\partial D_{t^0} \setminus \bar{U}$. Also, u is holomorphic at all points on $\partial D_{t^0} \cap \bar{U}$. Hence u extends to a neighborhood of $\bar{D}_{t^0}$, i.e., to D_s, for $s > t^0$ sufficiently close to t^0. □

REMARK 8.2. If all a_α, $|\alpha| = m$ are constants one can relax the hypothesis a bit to assume only that u is holomorphic in Ω_1. The argument proceeds along the following lines: Consider $\mathcal{O} \supset [z', z'']$, $\bar{\mathcal{O}} \subset \Omega_2$, where for $\mathcal{O}$ we can take a convex "tube" around $[z', z'']$.

CLAIM. There exists a compact set $K \subset \Omega_1$ such that all real hyperplanes H that meet $\mathcal{O}$ and are characteristic with respect to $\mathcal{L}$ also meet K.

PROOF OF THE CLAIM. Suppose there exists a sequence of points $z_j \in \bar{\mathcal{O}}$ and characteristic planes $H_j \ni z_j$ so that $H_j \cap K_j = \emptyset$ where $\{K_j\}$ is an increasing sequence of compact sets in Ω_1 satisfying $\bigcup_j K_j = \Omega_1$. By compactness we can assume that $z_j \to z^0 \in \bar{\mathcal{O}}$, $H_j \to H_0 \ni z^0$, H_0 being a characteristic plane. But then, $H_0 \cap K_j = \emptyset$ for all j, and so $H_0 \cap \Omega_1 = \emptyset$, a contradiction.

Now the remark follows from Corollary 8.1 if we take in place of Ω_1 a convex set U (with smooth boundary) such that $\bar{U} \subset \Omega_1$, $U \supset K \cup z'$, and in place of Ω_3 we take $\mathcal{O}$. □

3. Exercises

EXERCISE 8.1. Prove the following "Corollary" that follows rather from the method of proof than the statement of Theorem 8.1 (cf. Chapter 5).

COROLLARY 8.4. *Let $\mathcal{L}$ be a differential operator with constant coefficients in the principal part. Let $\Omega_1 \subset \Omega_2 \subset \mathbb{R}^n$ be bounded convex domains and all coefficients are real-analytic in $\bar{\Omega}_2$. Assume that for any $z^0 \in \Omega_2$ and any hyperplane $H \neq z^0$ characteristic with respect to $\mathcal{L}$, $H \cap \Omega_1 \neq \emptyset$. If u satisfies $\mathcal{L}u = 0$ in Ω_2 and $u \equiv 0$ in Ω_1, then $u \equiv 0$ in Ω_2.*

Hint: Repeat the construction in the proof of the Bony-Schapira theorem (modified as in Remark 8.2) and use Holmgren's theorem instead of Zerner's theorem.

EXERCISE 8.2. Prove the following:

COROLLARY 8.5 (cf. Theorem 7.1). *Let*

$$\mathcal{L} = \sum_{1}^{n} \frac{\partial^2}{\partial x_j^2} - \frac{\partial^2}{\partial x_{n+1}^2} + (\textit{lower order terms}),$$

where the coefficients appearing in the lower order terms are real analytic in $\mathbb{R}^{n+1}$. If a function $u \in C^2\left(\mathbb{R}^{n+1}\right)$ satisfies $\mathcal{L}u = 0$ and $u \equiv 0$ in a neighborhood of $\mathbb{R}^n \times \{0\}$, then $u \equiv 0$.

Hint: Choose Ω_1 and Ω_2 so that the hypothesis of the previous Corollary is satisfied.

EXERCISE 8.3. Prove that the constant $\frac{1}{\sqrt{2}}$ appearing in Corollary 8.2 is best possible. Namely, show that there is a function u harmonic in the ball of radius R in $\mathbb{R}^n$ that does not extend to a ball of radius larger than $\frac{R}{\sqrt{2}}$ in $\mathbb{C}^n$.
Hint: Consider a countable dense set S on the unit sphere $\{\|x\| = 1\}$ and a probability measure μ supported on S (an atomic measure). Take u to be the Poisson integral

$$u(y) = \frac{1}{C_n} \int_S \frac{1 - \|x\|^2}{\|x - y\|^n} d\mu(y),$$

where C_n is a normalizing constant (equal to the surface volume of the unit sphere, but this is not important). Show that u does not extend to any ball in $\mathbb{C}^n$ centered at the origin and of radius larger than $1/\sqrt{2}$.

Notes

Theorem 8.1 is due to J. Bony and P. Schapira [**22**]. Also, cf. [**86**]. Kiselman [**118**] proved a special version of the result for equations with constant coefficients by Fourier analysis methods. Corollary 8.2 is due to W. Hayman [**77**], whose proof is based on the power series expansion (also, cf. [**100**], [**178**]). More on this and other closely related topics can be found in the books by N. Aronszajn, Th. Creese and L. Lipkin [**10**] and V. Avanissian [**12**]. Corollary 8.4 is due to Hörmander [**86**, Theorem 8.6.5].

CHAPTER 9

Applications of the Bony-Schapira Theorem: Part I - Vekua Hulls

1. A uniqueness question for harmonic functions

The following simple question was posed by N. Nadirashvili [**144**] and addressed in [**115**] (cf. Q. 2 in Ch. 1). Consider the spherical shell $\Omega := \{x \in \mathbb{R}^n : r < |x| < R\}$, and let u be harmonic in Ω and vanish on the segment $(-R, -r)$ of, say, the x_1-axis (see Figure 1).
Q. 1: Does u also vanish on the remaining part of the x_1-axis, the segment (r, R)?

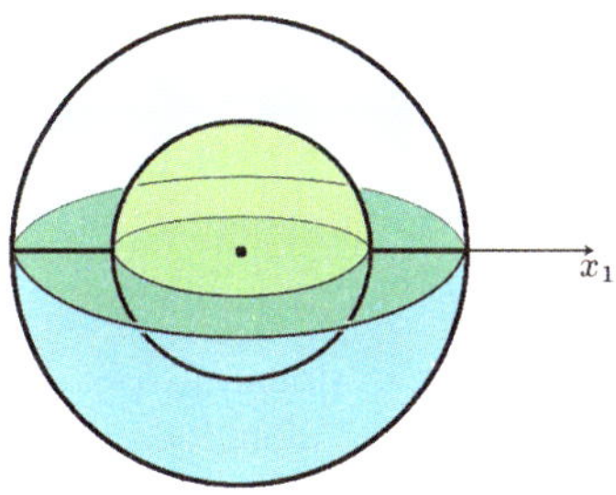

FIGURE 1. If a function u harmonic in a spherical shell Ω vanishes on the segment $(-R, -r)$, must u vanish on (r, R) as well?

For $n = 1$, Ω is disconnected and, hence, the answer is negative. Thus, we shall assume that $n \geq 2$. It is instructive to keep in mind the following simple example where a similar question has a negative answer.

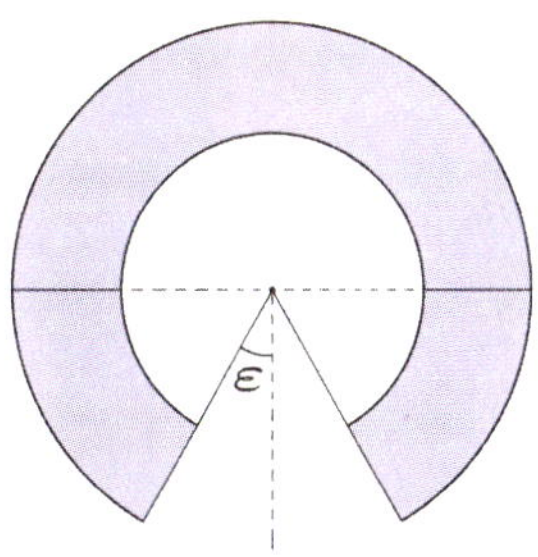

FIGURE 2. The function $u(z) = \arg z - \pi$ provides a negative answer in the slightly different case of an annulus with a sector removed.

EXAMPLE 9.1. Let $\Omega = \{z \in \mathbb{C} : r < |z| < R, -\frac{\pi}{2} + \varepsilon < \arg z < \frac{3\pi}{2} - \varepsilon\}$, where $\varepsilon > 0$. See Figure 2.

The function $u(z) = \arg z - \pi$ is harmonic in Ω and vanishes on the interval of the real axis $(-R, -r)$ but is equal to $-\pi$ on (r, R). This example illustrates that it is not enough for the domain to be connected, and if the answer to **Q. 1** is "Yes", then perhaps the symmetry plays a role.

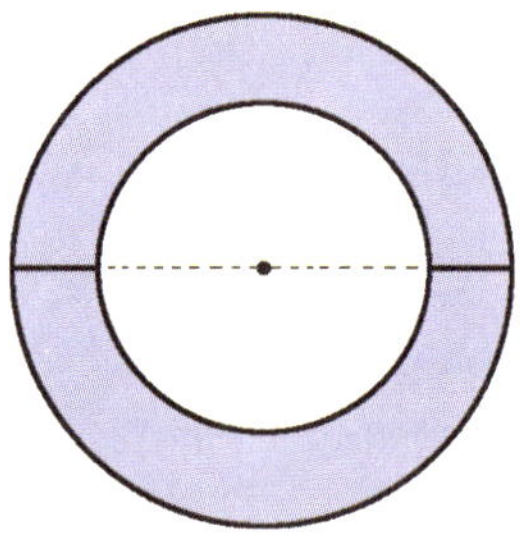

FIGURE 3. Exercise 9.1: Show that for $n = 2$ the answer to **Q. 1** is "Yes".

EXERCISE 9.1. Show using elementary complex analysis that for $n = 2$ the answer to **Q. 1** is "Yes". Hint: Consider the function $v(z) = u(z) + u(\bar{z})$, and apply the Schwarz reflection principle.

Thus inspired, one may ask further questions.

Q. 2: Suppose, even say in 2 dimensions, instead of the line $\{x_2 = 0\}$, we consider a line $\ell := \{x_2 = C\}$ (see Figure 4). Would the answer still be "yes"?

Q. 3: In the case of $n = 3$ dimensions, what happens if, instead of the spherical shell, we consider the intersection of the x_1-axis with a torus?

We wait to address this question until Section 3 since it is relevant to the discussion of the Conjecture 9.1 in that section; it turns out that the uniqueness property does not hold in the case of a torus.

Q. 4: What would be the answer to Q. 1 if we replace harmonic functions by polyharmonic functions, i.e. $u : \Delta^m u = 0$, m is an integer? More generally, what if u satisfies $\mathcal{L}u = 0$, $\mathcal{L} = \Delta^m + \sum_{|\alpha| \leq 2m-1} a_\alpha(x)\partial^\alpha$, where $a_\alpha(x)$ are, say, polynomials, or entire functions?

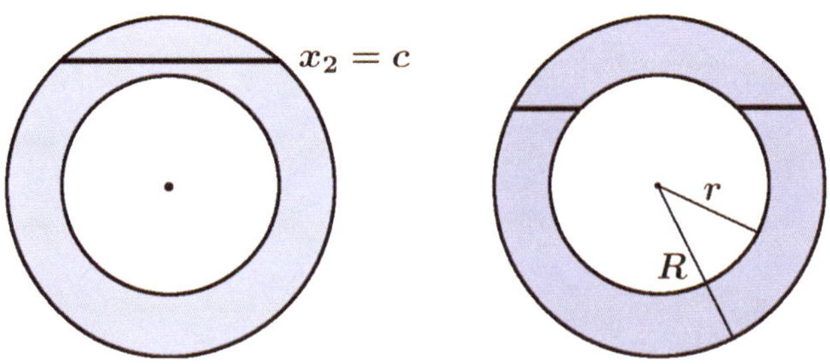

FIGURE 4. What if the line $\ell = \{x_2 = c\}$ does *not* pass through the center?

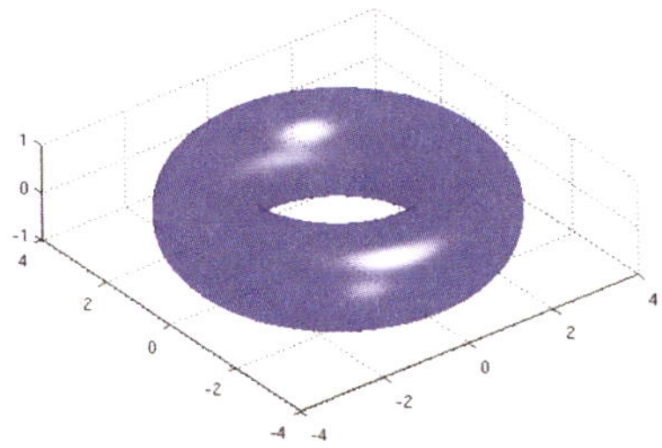

FIGURE 5. What if we consider a torus instead of a spherical shell?

For $R > |C| > r$, the intersection of ℓ and Ω is connected and the answer to Q. 2 is trivially "Yes". If $0 < |C| < r$, the question is, as we show below, fairly delicate and the answer depends on the "thickness" of the annulus.

Namely, if $R/r > 3$, the answer is "Yes" for all horizontal lines $x_2 = C$ for which $|C| < r$, and if $R/r \leq 3$ the answer is "Yes" if $0 < c < (R-r)/2$.

This result is sharp; if $R/r < 3$ and $(R-r)/2 < c < r$, the answer is "No" as demonstrated by an example due to T. Ransford; cf. [**115**]. We present his example in the following exercise.

EXERCISE 9.2. Let Ω an annulus in $\mathbb{C} \setminus \{0, \pm i\}$ which separates $\{0, -i\}$ from $\{i\}$. Define $u(z) = Re\sqrt{z(z-i)(z+i)}$. Check that u has a single valued branch on the annulus and that $u = 0$ on $\Omega \cap \{x_1 \leq 0\}$ and $u \neq 0$ on $\Omega \cap \{x_1 \geq 0\}$,

Note that the above annulus is "thin", i.e., the ratio of the radii $R/r < 3$ and can be easily made arbitrary close to 3.

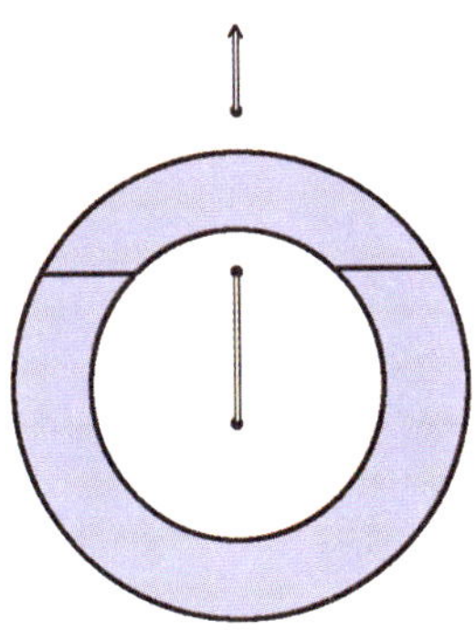

FIGURE 6. The branch cuts of $u(z) = Re\sqrt{z(z-i)(z+i)}$ are outside the annulus.

REMARK 9.1. It is worth noticing that due to Runge's theorem (the version for harmonic functions) there always exists a harmonic polynomial approximating an arbitrary function harmonic in a neighborhood of the union of the two segments $[-R, -r] \cup [r, R]$ obtained from "piercing" the spherical shell $\Omega := \{x \in \mathbb{R}^n : r < |x| < R\}$ by the x_1-axis. Indeed, to fix the ideas, let $n = 2$, and consider the function $v(x, y)$ harmonic in a (disjoint) neighborhood of the two segments and equal to 0 on the neighborhood of $[-R, -r]$ and 1 near $[r, R]$. By Runge's theorem, there exists an analytic polynomial $p(z)$, $z = x + iy$, such that $|p(z)| < \varepsilon$ near $[-R, -r]$ and $|1 - p(z)| < \varepsilon$ near $[r, R]$. Taking $u(x, y) = Re\ p$ we obtain a harmonic polynomial approximating 0 and 1 on $[-R, -r]$ and $[r, R]$ respectively within an arbitrary ε. The crux of the matter, of course, is that if we let $\varepsilon \to 0$ these harmonic polynomials may grow without bound somewhere in Ω away from the two intervals thus preventing them from converging to a harmonic function u in Ω vanishing on $[-R, -r]$ and equal to 1 on $[r, R]$.

EXERCISE 9.3. A modern proof of Runge's theorem resting on the Hahn-Banach theorem is quite streamlined and is the same in all dimensions (replacing analytic polynomials by harmonic polynomials).

Here is the sketch for the analytic case. We hope the reader will easily fill in the gaps.

(a) First note that it suffices to show that any function analytic in a neighborhood of a compact set K can be uniformly approximated on K by rational functions with poles off K. Then if the complement of K is connected the approximating rational functions can be replaced by polynomials.

(b) To prove the latter statement, suppose that a measure μ supported on K annihilates all rational functions with poles off K. If f is any function analytic in a neighborhood of K, we can represent it by the Cauchy integral over a contour γ surrounding K,

$$f(z) = \frac{1}{2\pi i} \int_\gamma \frac{f(\zeta)}{\zeta - z} d\zeta.$$

Since for every $\zeta \in \gamma$ and $z_0 \in K$, the function

$$\frac{1}{\zeta - z}$$

viewed as a rational function in z doesn't have poles on K, it is annihilated by μ, that is,

$$\int_K \frac{1}{\zeta - z} d\mu(z) = 0,$$

which implies

$$\int_\gamma \int_K \frac{f(\zeta)}{\zeta - z} d\mu(z) d\zeta = 0.$$

Changing the order of integration we obtain that

$$\int_K f d\mu = 0.$$

Applying the Hahn-Banach theorem then finishes the proof.

(c) Extend the above argument to $\mathbb{R}^n$ for $n > 2$, while replacing analytic functions by harmonic functions.

2. A view from $\mathbb{C}^n$: The Vekua hull

In this section we will explain what seems to be the high ground for the questions and examples from the previous section. Namely, given a domain Ω in $\mathbb{R}^n$, we will be interested in finding the maximal domain $\widehat{\Omega}$ in $\mathbb{C}^n$ to which all functions harmonic in Ω extend. Let us start out with the simple case when Ω is a disk in the plane.

EXAMPLE 9.2. The Lie ball in $\mathbb{C}^2$.

Let us find the maximal domain in $\mathbb{C}^2$ to which all functions harmonic in the unit disk $\mathbb{D} = \{(x,y) \in \mathbb{R}^2 : |x|^2 + |y|^2 < 1\}$ extend as holomorphic functions of complex variables X, Y.

PROPOSITION 9.1. *Any function u harmonic in $\mathbb{D}$ extends holomorphically to the Lie ball*

$$\widehat{\mathbb{D}} := \left\{ (X,Y) \in \mathbb{C}^2 : |X|^2 + |Y|^2 + 2\sqrt{x_2^2 y_1^2 + x_1^2 y_2^2 - 2x_1x_2y_1y_2} < 1 \right\}$$

where $X = x_1 + ix_2$ and $Y = y_1 + iy_2$.

PROOF. First note that $\Delta := \frac{\partial^2}{\partial x^2} + \frac{\partial^2}{\partial y^2} = 4\frac{\partial^2}{\partial z \partial \bar{z}}$, where recall that $\frac{\partial}{\partial z} := \frac{1}{2}\left(\frac{\partial}{\partial x} - i\frac{\partial}{\partial y}\right)$ and $\frac{\partial}{\partial \bar{z}} := \frac{1}{2}\left(\frac{\partial}{\partial x} + i\frac{\partial}{\partial y}\right)$. Solutions of $\frac{\partial^2}{\partial z \partial \bar{z}} u$ take the form $u = f(z) + g(\bar{z})$, where f, g are analytic in $\mathbb{D}$. Clearly, $f(z)$ extends as a holomorphic function in $\mathbb{C}^2$ to the cylinder $\{(X,Y) : X + iY \in \mathbb{D}\}$ by simply setting $f(X,Y) = f(X+iY)$. Similarly $g(\bar{z})$ extends as a holomorphic function to $\{(X,Y) : X - iY \in \mathbb{D}\}$, by $g(X,Y) = g(X - iY)$.

Thus u extends holomorphically to the domain in $\mathbb{C}^2$ defined by $\{(X,Y) : |X + iY| < 1 \text{ and } |X - iY| < 1\}$. Now

$$\begin{aligned} |X + iY|^2 < 1 &\Leftrightarrow (X + iY)\left(\bar{X} - i\bar{Y}\right) < 1 \\ &\Leftrightarrow |X|^2 + |Y|^2 + i\left(\bar{X}Y - X\bar{Y}\right) < 1 \\ &\Leftrightarrow |X|^2 + |Y|^2 - 2Im\left(X\bar{Y}\right) < 1. \end{aligned}$$

Similarly,

$$|X - iY|^2 < 1 \Leftrightarrow |X|^2 + |Y|^2 + 2Im\left(X\bar{Y}\right) < 1.$$

Thus, u extends to $\left\{(X,Y) : |X|^2 + |Y|^2 + 2\left|Im\left(X\bar{Y}\right)\right| < 1\right\}$ which is the Lie ball $\widehat{\mathbb{D}}$. □

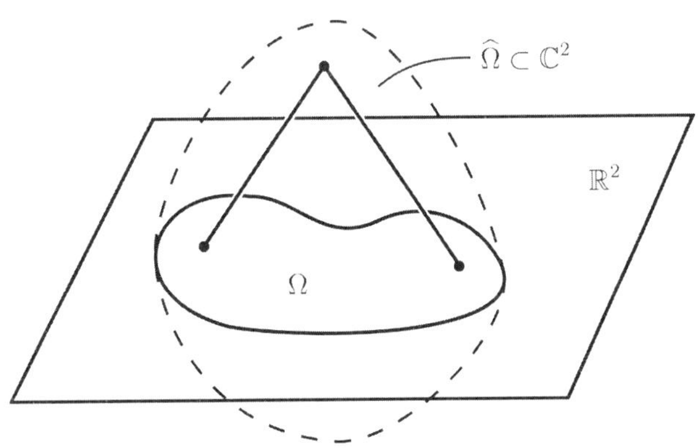

FIGURE 7. Any function harmonic in Ω extends to the Vekua hull $\widehat{\Omega}$.

Clearly this can be generalized to a domain Ω in the plane, and u a (complex) function harmonic in Ω. So $u = f(z) + g(\bar{z})$, where f and g are analytic in Ω and Ω^*, respectively. (Recall $\Omega^* = \{z : \bar{z} \in \Omega\}$.) Then, immediately f, g and, accordingly, u extend to a larger domain $\hat{\Omega}$ in $\mathbb{C}^2$, defined as

$$\hat{\Omega} := \{(X, Y) \in \mathbb{C}^2 : X + iY \in \Omega, X - iY \in \Omega^*\}.$$

DEFINITION 9.1. Let $\Omega \subset \mathbb{R}^2$ be a domain. A domain $\widehat{\Omega} \subset \mathbb{C}^2$ is called the Vekua hull (or, the harmonicity hull) of Ω if for any $(X, Y) \in \widehat{\Omega}$, $X + iY \in \Omega$, $X - iY \in \Omega^* := \{\bar{z} : z \in \Omega\}$.

EXAMPLE 9.3. Especially relevant to the uniqueness questions from the previous section, is the case when $\Omega = \{r < |z| < R\}$ is an annulus. Then the Vekua hull of Ω is $\hat{\Omega} = \{(X, Y) \in \mathbb{C}^2 : r < |X \pm iY| < R\}$.

Our uniqueness questions reduce to the following (See Figure 8):

Is the intersection of the complex line $\{Y = c\}$ with $\widehat{\Omega}$ connected?

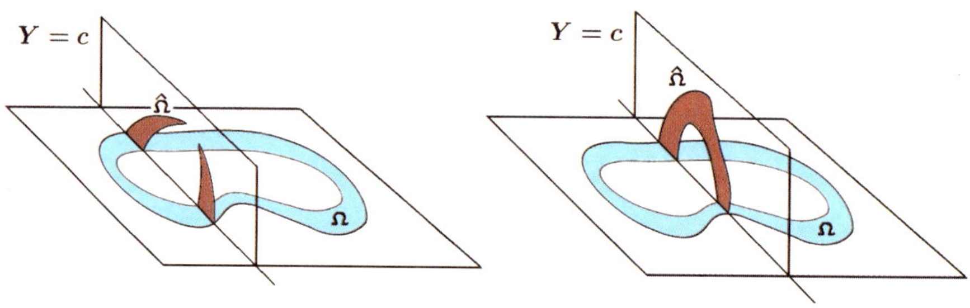

FIGURE 8. The intersection of $\hat{\Omega}$ with the complex line $\{Y = c\}$ can be disconnected or connected.

If it is, then u, hence being now a holomorphic function on $\{Y = c\} \cap \hat{\Omega}$ that vanishes on some part of this intersection, must vanish on the whole intersection! On the other hand, if the intersection $\{Y = c\} \cap \hat{\Omega}$ is disconnected, then there should be no reason for u to vanish on all the pieces of that intersection. However, we stress that this converse is, for the time being, just a conjecture – see Conjecture 9.1 near the end of this chapter.

Let us revisit Ransford's example, the "thin" annulus: $\Omega := \{r < |z| < R\}$, with $r < R \leq 3r$.

For each $c \geq 0$, the intersection $\{Y = c\} \cap \hat{\Omega}$ is an intersection of two annuli in the X variable: $\{(X, c) : r < |X \pm ic| < R\}$. This is a (real) 2-dimensional slice of $\hat{\Omega} \subset \mathbb{C}^2$. It is *disconnected* if $(R - r)/2 \leq c \leq r$, and *connected* if $c < (R - r)/2$ (See Figure 9).

In order to generalize this method, let us adapt slightly the concept of Globalizing Families (GF) from Chapter 5 to the situation studied in this section.

Let $V_1 \subset V_2 \subset \mathbb{C}^n$ be bounded open domains and

$$\mathcal{L} = \sum_{|\alpha| \leq m} a_\alpha \partial^\alpha$$

be a differential operator with coefficients a_α holomorphic in V_2.

We say that a family of domains $\{\Omega_t\}_{0 \leq t \leq 1}$ is a Globalizing Family (GF) for $(V_1, V_2, \mathcal{L})$ if the following properties hold:

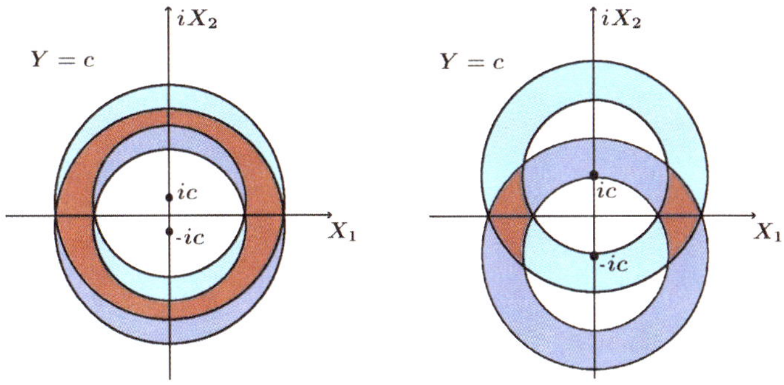

FIGURE 9. When Ω is an annulus, the set $\{Y = c\} \cap \hat{\Omega}$ is an intersection of two annuli in the X variable.

(i) All Ω_t are bounded and $\Omega_{t_1} \subset \Omega_{t_2}$, $t_1 < t_2$.
(ii) There exists $\delta > 0$ so that $\Omega_t \subset V_1$, $t < \delta$.
(iii) The family Ω_t is continuous, i.e.,

$$\Omega_{t_0} = \bigcup_{t<t_0} \Omega_t,$$

$$\overline{\Omega}_{t_0} = \bigcap_{t>t_0} \Omega_t.$$

(iv) All points on the boundaries of all $\partial\Omega_t \setminus V_1$ satisfy the hypothesis of Zerner's Theorem with respect to the operator $\mathcal{L}$. This entails that the boundaries $\partial\Omega_t \setminus V_1$ are sufficiently smooth (see Remark 4.1 following Zerner's Theorem).
(v) $V_2 = (\cup_t \Omega_t) \cup V_1$.

Then, we have the following "Meta-theorem" that we shall call the Globalizing Principle (GP).

GLOBALIZING PRINCIPLE 1. *If u is a holomorphic solution of $\mathcal{L}u = f$ in V_1 with f holomorphic in $\overline{V}_2$, then u extends holomorphically to V_2.*

(The proof of (GP) is immediate—cf. Chapter 5.) The following Corollary follows from the proof of the Bony-Schapira Theorem.

COROLLARY 9.1. *Let $\mathfrak{D}_1 \subset V \subset \mathbb{C}^n$ be bounded, open, convex domains. Suppose the coefficients of $\mathcal{L}$ are holomorphic in $\overline{V}$, and assume the Bony-Schapira condition holds, that is, every real characteristic hyperplane H that meets V must meet $\mathfrak{D}_1$ - cf. Chapter 8. Then for any $z^0 \in V$, there exists a convex domain $\mathfrak{D}_2 \subset V$ so that $z^0 \in \mathfrak{D}_2$ and a (GF) that extends any solution of $\mathcal{L}u = f$ holomorphic in $\overline{\mathfrak{D}_1}$ holomorphically to $\mathfrak{D}_2$ (provided that f is holomorphic in V).*

REMARK 9.2. If the principal part of $\mathcal{L}$ has constant coefficients, the same is true under the weaker assumption that $\mathcal{L}u = f$ in $\mathfrak{D}_1$ - cf. Remark 8.2 from the previous chapter.

We shall focus on the case $n = 2$ and

$$\mathcal{L} = \Delta^m + \sum_{|\alpha| \le 2m-1} a_\alpha(X, Y)\partial^\alpha.$$

For the sake of simplicity all the coefficients a_α and non-homogeneous terms are assumed to be entire functions in $\mathbb{C}^2$.

Let us observe the following property of the Vekua hull defined above in Definition 9.1.

PROPOSITION 9.2. *If Ω is convex, then $\widehat{\Omega}$ is also convex.*

PROOF. Let $Z_1 = (X_1, Y_1)$, $Z_2 = (X_2, Y_2) \in \widehat{\Omega}$. We want to show that $tZ_1 + (1-t)Z_2 \in \widehat{\Omega}$, $0 \leq t \leq 1$. We have

$$tX_1 + (1-t)X_2 + i(tY_1 + (1-t)Y_2 = t(X_1 + iY_1) + (1-t)(X_2 + iY_2) \in \Omega$$

because $X_j + iY_j \in \Omega$, $j = 1, 2$ and Ω is convex. Similarly,

$$tX_1 + (1-t)X_2 - i(tY_1 + (1-t)Y_2) = t(X_1 - iY_1) + (1-t)(X_2 - iY_2) \in \Omega^*$$

because Ω^* is also convex. □

COROLLARY 9.2. *Let $\Omega \subset \mathbb{C} \cong \mathbb{R}^2$ be a convex domain, $\widehat{\Omega}$ its Vekua hull. Then any solution of $\mathcal{L}u = f$ in Ω extends holomorphically to $\widehat{\Omega}$. Moreover, for any $z^0 \in \widehat{\Omega}$, there exists a convex domain $\mathfrak{D}_2$ containing z^0, $\mathfrak{D}_2 \subset \widehat{\Omega}$ and a (GF) extending solution u from $\mathfrak{D}_1$, a convex neighborhood of Ω in $\mathbb{C}^2$ to $\mathfrak{D}_2$.*

PROOF. In view of Definition 9.1 and Proposition 9.2 the corollary follows at once from Corollary 9.1. □

PROPOSITION 9.3. *Let Ω, $\Omega' \subset \mathbb{C}$ be simply connected domains. Let φ map Ω conformally onto Ω'. Then, there exists a biholomorphic map F of $\widehat{\Omega}$ onto $\widehat{\Omega'}$ (see Figure 10) so that $F|_\Omega = \varphi$ and F preserves all (GF) with respect to all the operators*

$$\mathcal{L} = \Delta^m + \sum_{|\alpha| \leq 2m-1} a_\alpha(z)\partial^\alpha, \quad m \geq 1,$$

with entire coefficients $a_\alpha(z)$.

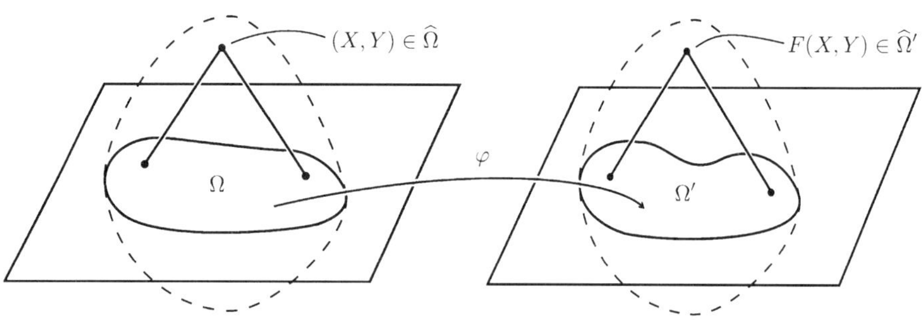

FIGURE 10. The conformal map $\varphi : \Omega \to \Omega'$ extends to a biholomorphism $F : \widehat{\Omega} \to \widehat{\Omega}'$.

PROOF. Changing the coordinates to $z = X + iY$, $w = X - iY$ we see that $\widehat{\Omega} \cong \Omega \times \Omega^*$, $\widehat{\Omega'} \cong \Omega' \times (\Omega')^*$. (Here, the notation $\cong$ means that the domains are biholomorphically equivalent, and, as before, $\Omega^* := \{\bar{z} : z \in \Omega\}$.) Setting $\varphi^*(w) = \overline{\varphi(\overline{w})}$, we notice that φ^* is a conformal mapping of Ω^* onto $(\Omega')^*$. Set

$F(z,w) := (\varphi(z), \varphi^*(w))$. Then F maps $\Omega \times \Omega^*$ onto $\Omega' \times (\Omega')^*$ biholomorphically. For the second part, we only have to check that F preserves non-characteristic points (with respect to $\mathcal{L}$) on smooth real hypersurfaces.

Let $\Gamma' := \{(z', w') : g(z', w') = 0\}$ be a C^1-real hypersurface in $\widehat{\Omega}'$ and let $\Gamma := \{(z,w) : G(z,w) = g(F(z,w)) = 0\}$ be its pull-back to $\widehat{\Omega}$. The characteristic points on Γ' are those satisfying

$$\frac{\partial g}{\partial z'} \frac{\partial g}{\partial w'} = 0$$

and on Γ:

$$\frac{\partial G}{\partial z} \frac{\partial G}{\partial w} = 0.$$

But $(F := (F_1, F_2))$,

$$\begin{aligned} \frac{\partial G}{\partial z} \frac{\partial G}{\partial w} &= \left(\frac{\partial g}{\partial z'} \frac{\partial F_1}{\partial z} \right) \left(\frac{\partial g}{\partial w'} \frac{\partial F_2}{\partial w} \right) \\ &= \frac{\partial g}{\partial z'} \frac{\partial g}{\partial w'} \varphi'(z) \left(\varphi^*\right)'(w) \end{aligned}$$

and the proposition follows since φ', $(\varphi^*)'$ do not vanish in Ω, Ω^*, respectively. □

COROLLARY 9.3. *Let u be a solution of $\mathcal{L}u = f$ in a simply connected domain $\Omega \subset \mathbb{C}$. Then, u extends holomorphically to $\widehat{\Omega}$.*

PROOF. Follows at once from Corollary 9.2, Proposition 9.3 and (GP) since by the Riemann mapping theorem (see [**6**, Ch. 6, Sec. 1]) Ω is conformally equivalent to a convex bounded domain Ω', e.g., the unit disk $\mathbb{D}$. □

COROLLARY 9.4. *Let as above $\mathcal{L} = \left(\frac{\partial^2}{\partial z \partial w}\right)^m +$ (lower order terms), and let $\Gamma = \{w = S(z)\}$ be a holomorphic hypersurface in $\mathbb{C}^2$. Suppose S is univalent in $U_1 \subset \mathbb{C}_z$, where U_1 is a simply connected and bounded domain. Let $U_2 = S(U_1)$. Then any solution of the Cauchy Problem*

$$\begin{cases} \mathcal{L}u = f; \\ \partial^\alpha (u - g) = 0, \ |\alpha| \le 2m - 1 \text{ on } \Gamma, \end{cases}$$

with f, g holomorphic in $\overline{\Omega} := \overline{U_1} \times \overline{U_2}$, extends holomorphically to Ω.

PROOF. Let ψ be a conformal mapping of U_2 onto the unit disk $\mathbb{D}$. Define $F(z,w) : \Omega \to \mathbb{D}^2$ by $F(z,w) = (\psi(S(z)), \psi(w))$. F is a biholomorphic isomorphism of Ω onto $\mathbb{D}^2$. Clearly,

$$F(\Gamma) = \left\{(z', w') \in \mathbb{D}^2 : z' = w'\right\} =: \Gamma'.$$

Γ' is convex and each real hyperplane through a point $Z' = \left(\left(z^0\right)', \left(w^0\right)'\right) \in \mathbb{D}^2$ that is characteristic with respect to the operator

$$\left(\frac{\partial^2}{\partial z' \partial w'}\right)^m + \text{(lower order terms)},$$

i.e., $Re\,(z') = Re\,\left(z^0\right)'$ or $Re\,(w') = Re\,\left(w^0\right)'$, intersects Γ'. Thus, for any convex neighborhood V' of Γ' and any point $Z' \in \mathbb{D}^2$ there is a (GF) with respect to $\left(V', \mathbb{D}^2, \mathcal{L}' := \left(\frac{\partial^2}{\partial z' \partial w'}\right)^m + \cdots\right)$ as in Corollary 9.1. The pre-image of this (GF) is then a (GF) with respect to a neighborhood V of Γ, Ω, $\mathcal{L}$ and the point $F^{-1}(Z') \in \Omega$. The corollary now follows from the (GP). □

REMARK 9.3. Let us pause for a moment to come back to question B posed in the Introduction but set in the dimension 2 rather than 3.

The function u harmonic in the annulus $\Omega = \{z \in \mathbb{C} : 1 < |z| < 2\}$ extends holomorphically to its Vekua hull $\widehat{\Omega}$ (perhaps, as a multivalued function, since Ω is not simply connected). It is easy to see that $\widehat{\Omega} \cap \{Y = 0\} \supseteq \{(X, 0) : 1 < |X| < 2\}$. Indeed, for any $(X^0, 0)$, $1 < |X^0| < 2$, both complex lines $X \pm iY = X^0$ intersect $\mathbb{R}^2$ at points $x^0 \pm iy^0 = X^0, \overline{X^0} \in \Omega$. Thus, the line segment $[-2, -1]$ and $[1, 2]$ that are disconnected in Ω are connected through $\widehat{\Omega}$. Hence, since u is holomorphic on $\widehat{\Omega} \cap \{Y = 0\}$ and vanishes on $\{(x, 0) : -2 < x < -1\}$ it must vanish on the whole intersection $\widehat{\Omega} \cap \{Y = 0\}$, in particular, on $\{(x, 0) : 1 < x < 2\}$. Although there is a simple direct proof of this fact for $n = 2$ (see Exercise 9.1), the advantage of this viewpoint is that with no extra work it allows the same conclusion for solutions of other equations, e.g., $\Delta^m u = 0$, or with a bit more care, even $\mathcal{L}u = 0$ with $\mathcal{L}$ as above. Still with a little more technicalities but essentially no new ideas, one can extend the result to higher dimensions to answer question B completely (see [**115**]).

The following exercise may help better familiarize the reader with the geometry of $\widehat{\mathbb{D}}$.

EXERCISE 9.4.

(i) Show that there exists a function u harmonic in $\mathbb{D}$ that does not extend holomorphically to any domain containing $\widehat{\mathbb{D}}$. Hint: See Exercise 8.3 at the end of the previous chapter.

(ii) Show that the n-dimensional analogue of the Lie ball is defined by

$$\widehat{B} := \left\{ z \in \mathbb{C}^n : \left(\|z\|^4 - \left| \sum_{j=1}^{n} z_j^2 \right|^2 \right)^{1/2} + \|z\|^2 < 1 \right\}.$$

In other words, $\widehat{B}$ is the largest domain in $\mathbb{C}^n$ to which all functions harmonic in the unit ball $B := \{x \in \mathbb{R}^n : \|x\| < 1\}$ extend as holomorphic functions. Show that the ball $B_{\mathbb{C}^n} = \{z \in \mathbb{C}^n : \|z\| < 1/\sqrt{2}\}$ is the largest ball centered at the origin contained in $\widehat{B}$.

(iii) Let $m \geq 1$. Show directly that any function u satisfying $\Delta^m u = 0$ in $\mathbb{D}$ extends holomorphically to $\widehat{\mathbb{D}}$. Hint: Prove that any such u can be written as

$$u(z) = u_0 + u_1|z|^2 + u_2|z|^4 + \cdots + u_{m-1}|z|^{2(m-1)},$$

where all u_j are harmonic in $\mathbb{D}$. (This is known as Almansi's expansion; see [**10**], [**12**].)

(Part (ii) of this exercise is not exactly routine and is intended for an ambitious reader. See Notes at the end of this chapter.)

3. Is the connectivity condition also necessary?

In the previous section, we observed that connectivity of the intersection of a complex line with the Vekua hull provides a simple sufficient condition for addressing the uniqueness questions stated at the beginning of the chapter. The following converse statement is known only in certain classes of examples, and especially *very* little is known in $n > 2$ dimensions.

CONJECTURE 9.1. *If a domain Ω intersects the x_1-axis in two components, then the uniqueness property for harmonic functions holds only if the Vekua hull intersected with the complex line $\{X \in \mathbb{C}^n : X_i = 0 \text{ for } i = 2, 3, ..., n\}$ is connected.*

The annulus and torus examples support the conjecture.

Let $\Omega = \left\{ \left(A - \sqrt{x_1^2 + x_2^2}\right)^2 + x_3^2 = a^2 \right\}$ be a torus in $\mathbb{R}^3$. Then, restating Q. 3, if a harmonic function $u(x)$ vanishes on the segment $[-A-a, -A+a]$ of the x_1-axis, then must $u(x)$ also vanish on the segment $[A-a, A+a]$?

The answer is "No" as we will now demonstrate by writing down a charge density supported in the exterior of the torus that generates a Newtonian potential constant on only one segment of the x_1-axis.

Without loss of generality, let $A = 1$, and let $u(x)$ be the Newtonian potential of the charge density

$$d\mu(x) = \sqrt{1 - (x_1 + 1)^2}ds$$

supported on the circle in the x_1x_3-plane $C := \{(x_1 + 1)^2 + x_3^2 = 1, x_2 = 0\}$ of radius 1 centered at $(-1, 0, 0)$. (ds denotes arc-length measure).

Then

$$u(x) = \int_C \frac{\sqrt{1 - (y_1 + 1)^2}ds}{|x - y|}.$$

Parametrize C using y_1, i.e., $y_3 = \pm\sqrt{1 - (y_1 + 1)^2}$. Then the arclength element is

$$ds = \sqrt{1 + \left(\frac{dy_3}{dy_1}\right)^2} = \sqrt{1 + \frac{(y_1 + 1)^2}{1 - (y_1 + 1)^2}} = \frac{1}{\sqrt{1 - (y_1 + 1)^2}}.$$

Taking $x = (x_1, 0, 0)$, we have

$$\begin{aligned} u(x_1, 0, 0) &= 2\int_{-2}^{0} \frac{dy_1}{\sqrt{(x_1 - y_1)^2 + 1 - (y_1 + 1)^2}} \\ &= 2\int_{-2}^{0} \frac{dt}{\sqrt{x_1^2 - 2x_1t - 2t}} \\ &= \frac{2}{-1 - x_1}\left(\sqrt{x_1^2} - \sqrt{x_1^2 + 4x_1 + 4}\right) \\ &= \frac{2}{-1 - x_1}\left(|x_1| - |x_1 + 2|\right). \end{aligned}$$

This last expression is constant $= 4$ for $-2 < x_1 < 0$, and it is monotone decreasing for $x_1 > 0$.

EXERCISE 9.5 (Alternative torus example.). Using the mass density x_3^2 on the same ring, show that the force vanishes along the left segment on the x_1-axis. (This implies that the potential is constant along the left segment, and it is obviously decreasing along the right segment.)

Hint: Take a point $M = (x_1, 0, 0)$ on that left segment. Consider the attraction force at M from two points P, Q where the chord through M hits the circle.

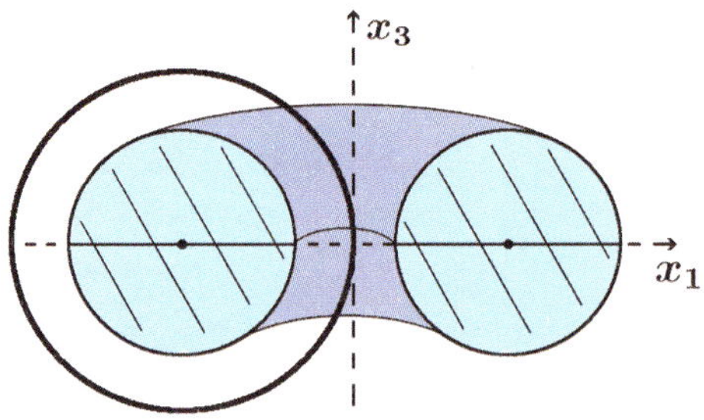

FIGURE 11. Cross-sectional view: a carefully chosen mass density supported on a circle generates a Newtonian potential that is constant on only one segment of the x_1-axis intersecting the torus.

Introducing segments from P and Q perpendicular to the x_1-axis, use two similar right triangles to show that the force at M due to P and Q cancels. Summing the force at every such pair of points (in fact taking infinitesimal arcs), conclude that the total force at M vanishes.

EXERCISE 9.6. Show that for Ω equal to the torus from the previous section, the intersection of $\hat{\Omega}$ with the x_1-axis is disconnected.

Perhaps if the torus example and the thin annulus example can be understood from the high ground $\mathbb{C}^n$ point of view it might lead to a proof of Conjecture 9.1.

Notes

Solution to Exercise 9.1: By the reflection principle, near the segment $(-R, -r)$, u is an odd function with respect to y. Hence, $u(z)+u(\bar{z}) = 0$. Since $\bar{z} \in \Omega$ whenever $z \in \Omega$, this equation is valid throughout Ω including at the segment (r, R).

The notion of the Vekua hull (of course, not the name) appears in Vekua's work [**196**]. Also, cf. [**64**], [**81**] and the references therein. The extension to higher dimensions under the name of "hull of harmonicity" is due independently to P. Lelong and N. Aronszajn—cf. [**10**], [**12**]. The Lie ball is discussed in [**10**], [**12**] in great detail. Corollaries 9.3 and 9.4, the main achievements of Vekua's theory, are due to Vekua [**196**], [**81**]. His methods are based on explicit representations of solutions by Riemann's formulae for solutions of hyperbolic equations in two variables and restricted to second order operators, or constant coefficient operators, for the higher order case. Also, Vekua states Corollary 9.4 for a domain $\Omega \subset \mathbb{R}^2$ with the Cauchy problem posed on an analytic curve $\gamma \subset \Omega$. In that context, one has to assume that Ω is conformally symmetric with respect to γ (cf. [**196**], [**31**], [**81**]). That means that γ is given by the equation $\bar{z} = S(z)$, S being an analytic function in Ω so that $\overline{S}$ maps Ω bijectively onto itself. (More on this is contained in Chapter 11). The presentation here, based on a powerful though geometrically simple method of Globalizing Families allowing to bypass entirely the explicit representation formulae, and essentially ignore the lower order terms of the operator, is from [**112**].

CHAPTER 10

Applications of the Bony-Schapira Theorem: Part II - Szegő's Theorem Revisited

Let us discuss here a remarkable theorem of Szegő [**186**] on locating the singularities of axially symmetric harmonic functions in $\mathbb{R}^3$ (cf. the question raised in Sec. 4 of Ch. 1). Recall first some standard facts regarding spherical harmonics (cf., e.g. [**99**]). Let r, θ, ϕ denote spherical coordinates of a generic point $p \in \mathbb{R}^3$, where $0 \leq \theta \leq \pi$ is the angle measured from the positive x_3-axis to the vector joining the origin to the point $p \in \mathbb{R}^3$ (see Figure 1), and $0 \leq \phi \leq 2\pi$ is the (azimuthal) angle measured from the x_1 axis to the projection of p to the x_1x_2 - plane.

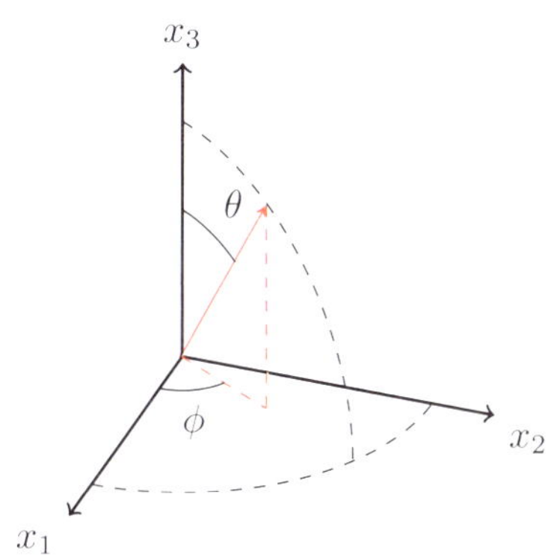

FIGURE 1. Spherical coordinates.

Let now u be a real-valued harmonic function in the open unit ball B which is axially symmetric about the x_3-axis, i.e., independent of ϕ. In other words it is of the form $u(x_1, x_2, x_3) = f(x_1^2 + x_2^2, x_3)$. As is well-known (see for example, [**99**]), u admits the expansion

$$u = \sum_{n=0}^{\infty} a_n r^n P_n(\cos\theta), \tag{10.1}$$

where P_n denotes the Legendre polynomial of degree n and a_n are real constants satisfying

$$\limsup |a_n|^{1/n} \leq 1. \tag{10.2}$$

(The functions $r^n P_n(\cos\theta)$ are called zonal harmonics [**99, 187**]. Suppose now that

$$\limsup |a_n|^{1/n} = 1, \tag{10.3}$$

so that (10.1) converges in B and no larger concentric ball. Then u must have singularities on the unit sphere, and it is of interest to locate them. Note that if $p \in \partial B$ is a singularity, the whole circle of latitude containing p is also singular. The problem is then to find a criterion for determining the regularity or singularity of a given circle of latitude.

Thus, in short, let $u = \sum_{n=0}^{\infty} a_n r^n P_n(\cos\theta)$ be an axially symmetric harmonic function in the open unit ball, where P_n are Legendre polynomials, and $a_n \in \mathbb{R}$, with

$$\overline{\lim} |a_n|^{1/n} = 1.$$

This implies that u has singularities on the unit sphere, but where?

The remarkable theorem of Szegő pinpoints precisely where the singularities occur.

THEOREM 10.1 ([**186**]). *The function $u = \sum_{n=0}^{\infty} a_n r^n P_n(\cos\theta)$ extends harmonically across $p_0 = (1, \theta_0, \phi)$ if and only if the Taylor series $f(\zeta)=\sum_{n=0}^{\infty} a_n \zeta^n$ extends holomorphically across $\zeta_0 = e^{i\theta_0}$.*

This is a useful result because it reduces the study of the harmonic continuation of the (previously) mysterious zonal harmonic expansion (10.1) to the well-studied holomorphic continuation for Taylor series, about which there is a great wealth of results (see, for example [**21**], [**33**]). For example, as Szegő remarked, by choosing $\{a_n\}$ so that $a_n = 0$ on a set of integers with density 1 (in which case $f(\xi) = \sum_{n=0}^{\infty} a_n \xi^n$ in Theorem 10.1 is not continuable across any point of the unit circle by Fabry's well-known theorem), we get examples of expansions (10.1) not continuable across any point of the unit sphere (hence, a fortiori, harmonic functions in B not continuable across any boundary point). Also, the analogous result for k-dimensional zonal harmonic expansions (whereby, in place of the Legendre polynomials appearing in (10.1), we have ultraspherical ones of appropriate order) is also true, as we explain below.

1. Jacobi polynomial expansions: Generalization of Szegő's theorem

Let us reformulate the problem in a more general setting. Change variables $r = e^x$, $x < 0$, $\theta = y$. So

$$v_1(x, y) = \sum_{n=0}^{\infty} a_n e^{nx} P_n(\cos y)$$

is real analytic in the left half-plane.

Notice that

$$Re\left(\sum_{n=0}^{\infty} a_n \zeta^n\right) = \sum_{n=0}^{\infty} a_n r^n \cos n\theta = \sum_{n=0}^{\infty} a_n r^n T_n(\cos\theta),$$

where T_n denotes the Chebyshev polynomial of the first kind.

Changing variables as before we obtain the function

$$v_2(x, y) = \sum_{n=0}^{\infty} a_n e^{nx} T_n(\cos y),$$

which is real-analytic in the left half-plane.

Assuming henceforth that (10.3) holds $v_1(x,y)$ is a real-analytic function in the left half-plane $\{x<0\}$ and the Szegő problem is now to find its singularities on the y-axis (because of the 2π-periodicity in y, we can restrict our attention to the strip $-\pi \le y \le \pi$).

Szegő's Theorem can then be reformulated as locating the singularities (on the y-axis) of v_1 in terms of those of v_2. But each of $\{P_n\}$ and $\{T_n\}$ is just a single specimen from the vast hierarchy of Jacobi orthogonal polynomials $P_n^{\alpha,\beta}$, $\alpha,\beta>-1$. $\alpha=\beta=0$ correspond to Legendre polynomials, and $\alpha=\beta=-1/2$, Chebyshev polynomials. More generally, $\alpha=\beta=\frac{k-3}{2}$ correspond to general ultraspherical polynomials appearing in zonal harmonic expansions in $\mathbb{R}^k$, $k\ge 3$. (We refer the reader to Szegő's classic [**187**, Ch. 4] for the background on various classes of Jacobi polynomials.) This suggests a possible generalization of Szegő's theorem.

THEOREM 10.2 ([**42**]). *Suppose*

$$\overline{\lim}|a_n|^{1/n}=1,$$

so that for all $\alpha,\beta>-1$ the series

$$u_{\alpha,\beta}(x,y)=\sum_{n=0}^{\infty} a_n e^{nx} P_n^{\alpha,\beta}(\cos y) \tag{10.4}$$

defines a real-analytic function in $\{x<0\}$. If for a given $y_0\in\mathbb{R}$, $u_{\alpha,\beta}(x,y)$ extends as a real-analytic function across $(0,y_0)$ for some $\alpha,\beta>-1$, then the same is true for all $u_{\alpha,\beta}$, with any $\alpha,\beta>-1$.

In terms of the original variables r and θ, the theorem states that for *any* $\alpha,\beta>-1$ the function $u_{\alpha,\beta}$ is continuable across $(1,\cos\theta_0)$ if and only if the Taylor series $f(\zeta)=\sum_{n=0}^{\infty} a_n\zeta^n$ is analytically continuable across $e^{i\theta_0}$.

REMARK 10.1. Theorem 10.2 is known in principle in closely related settings, see for instance [**199**, **202**], where quite different methods are involved.

REMARK 10.2. The proof outlined below takes the high ground point of view developed in the previous chapters. Szegő's proof, while simple and ingenious, is not related to propagation of singularities for analytic PDE.

REMARK 10.3. From the proof below, it will also be clear that one can make some assertions concerning simultaneous global continuation of $u_{\alpha,\beta}$ based on the techniques developed in the previous chapters.

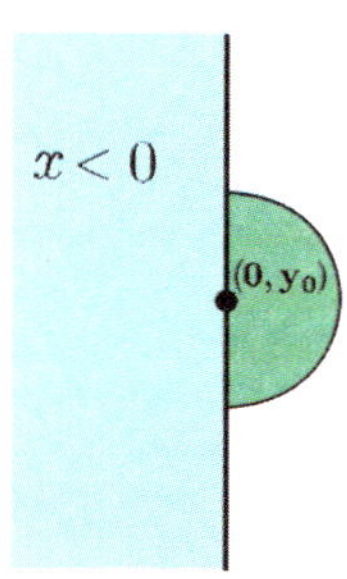

FIGURE 2. In the new variables (x,y), the problem is to extend $u_{\alpha,\beta}$ from the left halfplane $\{x<0\}$ across a point $(0,y_0)$.

2. Relation to holomorphic PDEs

First of all, observe that all $u_{\alpha,\beta}(x,y)$ satisfy the PDE

$$M_{\alpha,\beta}u_{\alpha,\beta}(x,y) = 0, \tag{10.5}$$

where

$$M_{\alpha,\beta} = \Delta_{x,y} + (\alpha+\beta+1)\frac{\partial}{\partial x} + \frac{\alpha-\beta+(\alpha+\beta+1)\cos y}{\sin y}\frac{\partial}{\partial y}.$$

To derive this PDE, let us write $w = w(t) = P_n^{\alpha,\beta}(t)$ and start from the well-known ODE satisfied by the Jacobi polynomials

$$(1-t^2)w'' + ((\beta-\alpha)-(\alpha+\beta+1)t)w' + n(n+\alpha+\beta+1)w = 0. \tag{10.6}$$

Letting, for fixed α, β,

$$f_n(y) = P_n^{\alpha,\beta}(\cos y) = w(\cos y),$$

we easily calculate

$$w'(\cos y) = \frac{f_n'(y)}{\sin y}, \tag{10.7}$$

$$w''(\cos y) = \frac{\sin y f_n''(y) - \cos y f_n'(y)}{\sin^3 y}. \tag{10.8}$$

Substituting $t = \cos y$ into (10.6) and using (10.7 - 10.8) we obtain, after some simplifications,

$$f_n''(y) + \frac{\beta-\alpha+(\alpha+\beta+1)\cos y}{\sin y}f_n'(y) + n(n+\alpha+\beta+1)f_n(y) = 0. \tag{10.9}$$

Let us define a differential operator ℓ by

$$\ell f = f'' + \frac{\alpha-\beta+(\alpha+\beta+1)\cos y}{\sin y}f', \tag{10.10}$$

in order to rewrite (10.9) as

$$\ell f = -\lambda_n f', \quad \lambda_n := n(n+\alpha+\beta+1). \tag{10.11}$$

Letting now L denote the differential operator

$$Lg = g'' + (\alpha+\beta+1)g', \tag{10.12}$$

we see that (where the subscripts x, y indicate the variable of differentiation)

$$L_x(e^{nx}f_n(y)) = \lambda_n e^{nx}f_n(y), \tag{10.13}$$

and

$$\ell_y(e^{nx}f_n(y)) = -\lambda_n e^{nx}f_n(y), \tag{10.14}$$

so that $e^{nx}f_n(y) = e^{nx}P_n^{\alpha,\beta}(\cos y)$ is annihilated by $L_x + \ell_y$, which is the same partial differential operator $M_{\alpha,\beta} = L_x + \ell_y$ that appears in (10.5). Since $M_{\alpha,\beta}$ is linear and independent of n, it also annihilates $u_{\alpha,\beta}$ defined by (10.4), our first goal is thus achieved. We summarize it as follows.

LEMMA 10.1. *The functions $u_{\alpha,\beta}(x,y) := \sum a_n e^{nx}P_n^{\alpha,\beta}(\cos y)$ satisfy a family of PDEs (10.5) that all have the same principal part.*

3. Proof of the generalized Szegő theorem

We shall carry out the argument for proving Theorem 10.2 in two parts.

(a) First, we assume that for some (α, β) the function $u_{\alpha,\beta}$ continues across $(0, y_0)$ and show that the corresponding Chebyshev expansion (the case $\alpha = \beta = -1/2$) also extends, that is the harmonic function

$$h(x, y) = \sum_{n=0}^{\infty} a_n e^{nx} \cos ny \tag{10.15}$$

continues across $(0, y_0)$.

(b) Conversely, assuming the continuability of h we shall deduce the continuability of $u_{\alpha,\beta}$ for all α, β.

Let us sketch the argument for (a). So, suppose that $u_{\alpha,\beta}$ continues across $(0, y_0)$. By virtue of periodicity and evenness of $u_{\alpha,\beta}(x, y)$ with respect to y, without loss of generality we can assume that $0 \le y \le \pi$. Then $u_{\alpha,\beta}$ also continues across $(0, -y_0)$. Hence we can construct a simply-connected domain U (see Fig. 3), symmetric with respect to the x-axis, containing $(0, \pm y_0)$ and such that $u_{\alpha,\beta}$ is real-analytically extendable to all of U. Now here comes the main "punch" that requires an extension of Zerner's theorem that is the main result in [**42**]. $u_{\alpha,\beta}$ satisfies the PDE as in Lemma 10.1 with the Laplacian in the principal part, yet the lower order coefficients of $M_{\alpha,\beta}$ have a simple "polar" singularity on the complex line $\{y = 0\}$. Imagine for a moment that $M_{\alpha,\beta}$ has in fact smooth coefficients. Then using the method of globalizing families as in the previous chapter, we can extend $u_{\alpha,\beta}$ to the Vekua hull $\hat{U}$ of U. This implies that the Cauchy data for $u_{\alpha,\beta}$ on $\{y = 0\}$ is holomorphic near $\hat{U} \cap \{y = 0\}$. However, $u_{\alpha,\beta}(x, y)$ shares the Cauchy data on $\{y = 0\}$ with a harmonic function

$$k(x, y) = \sum_{n=0}^{\infty} a_n P_n^{\alpha,\beta}(1) e^{nx} \cos ny. \tag{10.16}$$

This latter series converges for $\{x < 0\}$ since the nth root of the factor $P_n^{\alpha,\beta}(1) = \binom{n+\alpha}{n}$ converges to 1.

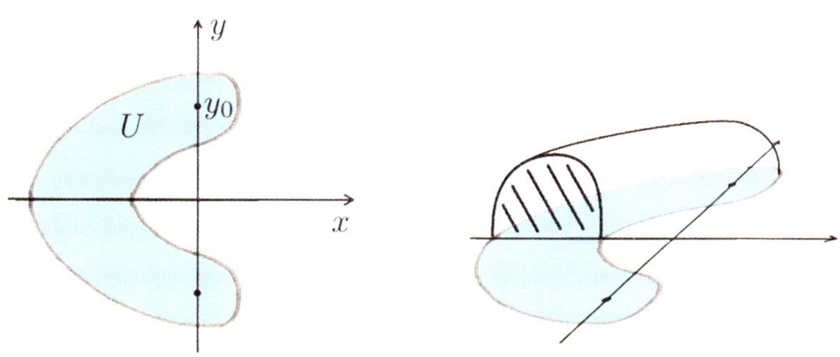

FIGURE 3. We can extend first extend $u_{\alpha,\beta}$ first to a domain U symmetric with respect to the x axis (left) and then to the Vekua hull $\hat{U}$ of U (right).

Now, it is geometrically obvious that the Vekua hull of $\hat{U} \cap \{Y = 0\}$ contains U. To see that it only suffices to observe that all bicharacteristic lines $\{X \pm iY = C\}$

emanating from points in $\hat{U}\cap\{Y=0\}$ must intersect U by the very definition of the Vekua hull. Hence invoking results from Chapter 9, we can construct a globalizing family that extends $k(x,y)$ from $\hat{U}\cap\{Y=0\}$ to all of $\hat{U}$, in particular, to $(0,\pm y_0)$. It remains to deduce from this that $h(x,y)$ also extends across $\hat{U}\cap\{Y=0\}$. In terms of the variables used earlier, this amounts to showing that if $\sum_{n=0}^{\infty} a_n P_n^{\alpha,\beta}(1)\zeta^n$ is analytically continuable across e^{iy_0} the same is true for $\sum_{n=0}^{\infty} a_n\zeta^n$. This latter statement follows from Hadamard's "multiplication of singularities" theorem once one shows that

$$\sum_{n=0}^{\infty} P_n^{\alpha,\beta}(1)^{-1}\zeta^n \tag{10.17}$$

(which converges in the open unit disk $\mathbb{D}$) is analytically continuable across $\partial\mathbb{D}\setminus\{1\}$. This follows from a well-known theorem of Lindelöf [**21**, Satz 7.3.VIII] (cf. [**33**]) since the function $A(n)$ given by the nth coefficient of (10.17)

$$A(n) = P_n^{\alpha,\beta}(1)^{-1} = \frac{\Gamma(n+1)\Gamma(\alpha+1)}{\Gamma(n+\alpha+1)}$$

extends as an analytic function with subexponential growth in the right halfplane.

Now to make this above plan work requires an extension of Zerner's theorem at the points $\{Y=0\}\cap\hat{U}$ where the globalizing family will have characteristic points if we replace the operator $M_{\alpha,\beta}$ by $yM_{\alpha,\beta}$. The following result from [**42**] provides the desired extension of Zerner's theorem. The statement in [**42**, Thm. 1.2] includes more general operators, but in order to keep the presentation as simple as possible we will only state a specific case of the result adapted to our situation.

THEOREM 10.3. *Let u satisfy $P(D)u=0$ in a domain $\Omega\subset\mathbb{C}^2$ with smooth boundary, with $P(D)=y\Delta+$ lower order terms and where all the coefficients of $P(D)$ are holomorphic in $\overline{\Omega}$. Assume that $\partial\Omega$ is non-characteristic with respect to Δ at points of $\partial\Omega\cap\{Y=0\}$ and that the complex line $\{Y=0\}$ meets $\partial\Omega$ transversally. Then, u extends holomorphically to a neighborhood of $\partial\Omega\cap\{Y=0\}$.*

Applying Theorem 10.3 to a globalizing family sweeping out $\hat{U}$ completes the first part of the extension of Szegő's theorem.

Before discussing the ideas behind Thereom 10.3 let us finish our outline of the extended Szegő's theorem by addressing point (b). Thus, assume that the function (10.15) extends across $(0,\pm y_0)$. Once again, invoking Hadamard's multiplication of singularities theorem we conclude that the same is true for all functions $k(x,y)=k_{\alpha,\beta}(x,y)$ as defined in (10.16). All these functions are harmonic, and by a globalizing family argument they all extend to $\hat{U}$, the Vekua hull of the domain U. Therefore, their Cauchy data on $\{Y=0\}$ are all holomorphic near $\hat{U}\cap\{Y=0\}$. But the latter are the same as for $u_{\alpha,\beta}$. Now, the operator $yM_{\alpha,\beta}$ doesn't satisfy the Cauchy-Kovalevskaya theorem near $\{Y=0\}$ since that plane is obviously characteristic for it. Yet, the "mild" nature of the singularity in the lower order coefficient of $M_{\alpha,\beta}$ allows for an extended version of the Cauchy-Kovalevskaya theorem to apply [**140**]. (The situation is similar to that of "Fuchsian" singularities for ODE.) Consequently, all $u_{\alpha,\beta}$ are analytic at least in a neighborhood of $\hat{U}\cap\{Y=0\}$. But this takes care of all "bad" points on the boundaries of domains in the globalizing family extending solutions of analytic PDE with the Laplacian in the principal part from $\hat{U}\cap\{Y=0\}$ to $\hat{U}$. Thus, all $u_{\alpha,\beta}$ continue analytically to U and we have (b).

Let us sketch now the main idea behind Theorem 10.3. To keep it simple, let us assume the boundary of Ω is given by $\{z \in \mathbb{C}^2 : ImF(z) = 0\}$, where F is a holomorphic function with non-vanishing gradient near, say, 0 and $F(0) = 0$. Then, we want to show that any holomorphic solution of

$$Y\Delta u + (\text{lower order terms}) = f,$$

where f is holomorphic in $\overline{\Omega}$, extends across 0. Without loss of generality, $\Omega = \{ImF(z) < 0\}$, and $S = \{Y = 0\}$ meets $\{ImF(z) = 0\}$ transversally at 0. Zerner's theorem allows us to extend u to a larger domain Ω' across all points $z^1 \in \partial\Omega \setminus S$.

Let Γ be an analytic disk in $\partial\Omega$ (in some small neighborhood of 0) that meets S only at the origin. Thus, all points $p \neq 0$ on Γ are non-characteristic points of $\partial\Omega$, and hence u can be continued across every such point by Zerner's theorem. We may therefore assume that u is holomorphic in some larger set Ω' containing Ω and also containing $\Gamma \setminus \{0\}$, with 0 being on the boundary of Ω'. By moving Γ slightly into Ω we obtain a family of analytic disks contained in Ω' with a limiting analytic disk that meets the boundary only at 0. For example, if $\partial\Omega = \{ImX = 0\}$, we can take for Γ an analytic disk inside the complex line $\{X = 0\}$, apply Zerner's theorem across every point (except the origin) on that line obtaining a larger domain Ω' to which the solution u extends. Note that $\partial\Omega'$ looks like a "saddle" near the origin. Now construct a family of analytic disks Γ_ε choosing analytic disks contained in the complex lines $\{X = -i\varepsilon\}$. Note that the limiting analytic disk meets the boundary of Ω' only at the origin that is an interior point for that analytic disk - cf. Figure 4. By the Behnke-Sommer continuity principle [**87**, Lemma 2.5.3] all holomorphic functions in Ω' (in particular, u) extend holomorphically across 0. This completes the proof of Theorem 10.3 in this simple situation. The general case, although quite technical, does not invoke any novel ideas. The outline of the PDE outlook on Szegő's theorem is now complete.

Note that in using the Behnke-Sommer continuity principle (a.k.a., "Kontinuitätsatz"), we invoke a phenomenon of forced analytic continuation coming from the general theory of several complex variables, whereas Zerner's theorem is an analytic continuation principle specifically adapted to solutions of holomorphic PDE. The two principles complement each other rather serendipitously in the particular situation above.

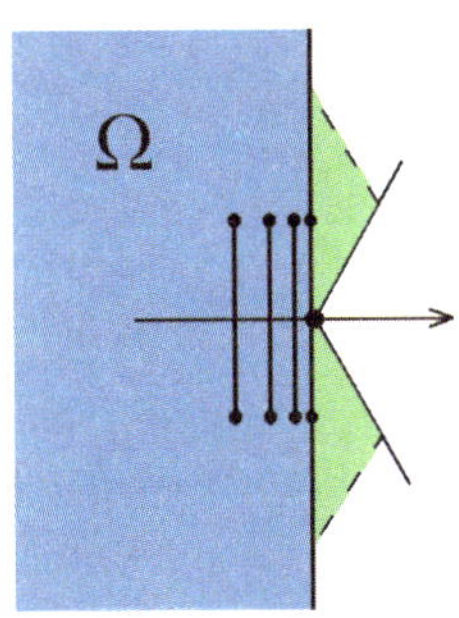

FIGURE 4. The extension Ω' of the domain Ω admits a family of analytic disks with a limiting disk that meets the boundary of Ω' only at the origin. (Note: the analytic disks are depicted here as line segments, and Ω' consists of Ω along with the triangular regions.)

REMARK 10.4. Essentially the same argument shows that if, for some α, β, the function $u_{\alpha,\beta}$ defined in Lemma 10.1 is continuable from the left halfplane along a horizontal segment to where $x_0 > 0$ then the same is true for any other choice of α, β. Indeed, we have only to construct our domain U so that, in place of $(0, y_0) \in U$, we have $(x_0, y_0) \in U$; this is easily achieved by letting U be the union of $\{x < 0\}$ and neighborhood of two closed segments one joining $(0, y_0)$ to (x_0, y_0) and the other joining $(0, -y_0)$ to (x_0, y_0). (Of course, if $y_0 = 0$ only one segment is needed, and U can be made bounded by intersecting with a sufficiently large disk.) In terms of the original variables this allows us to state the result as: *for all choices α, β, the Mittag-Leffler star of real-analyticity of $\sum_{n=0}^{\infty} a_n r^n P_n^{\alpha,\beta}(\cos\theta)$ is the same, thus allowing us a "free" global analytic continuation statement for all zonal harmonic expansions.* (Here, we define the Mittag-Leffler star in word for word analogy with the classical Taylor series situation; cf. [**21**, **33**].)

EXERCISE 10.1. Verify Szegő's Theorem in the case of the Newtonian potential of a unit charge (mass) located at the North Pole of the unit sphere in $\mathbb{R}^3$, by sorting out the singularities of the Legendre polynomial expansion.

EXERCISE 10.2. Use Szegő's Theorem in order to provide an example of an axially symmetric harmonic function that cannot be analytically continued past any point on the unit sphere.

4. Nehari's theorem revisited

Inspired by Szegő's theorem Nehari in 1956 [**145**] expanded it to a new setting to deal with analytic continuation of expansions of Legendre polynomials in an ellipse, a large step from the classical setting of the Taylor series in the disk. To fix the ideas let $\{P_n(t)\}$ denote the classical Legendre polynomials orthogonal on $[-1, 1]$ and normalized by $P_n(1) = 1$ as before (cf. [**186**]). Let $\{a_n\}_{n=0}^{\infty}$ be a sequence of complex numbers satisfying

$$\limsup |a_n|^{1/n} = 1/R < 1. \tag{10.18}$$

Then, as is well-known, for each complex $t \notin I := [-1, 1]$ the following asymptotic formula holds.

$$\lim_{n\to\infty} |P_n^{\alpha,\beta}(t)|^{1/n} = \varphi(t), \tag{10.19}$$

where

$$\varphi(t) := t + (t^2 - 1)^{1/2}, \tag{10.20}$$

the branch of the square root being taken so that $\varphi(t) \to \infty$ along the positive real axis (cf. [**187**, formula (8.23.1)]). In fact, (10.19) holds for all Jacobi polynomials $P_n^{\alpha,\beta}, \alpha, \beta > -1$. The function φ maps $\mathbb{C} \setminus I$ conformally on $\{s \in \mathbb{C} : |s| > 1\}$ so that $\varphi(\infty) = \infty$. We denote then the inverse map by $t = \psi(s)$, where

$$\psi(s) := \frac{1}{2}\left(s + s^{-1}\right), \quad |s| > 1. \tag{10.21}$$

The set

$$E_r := \{t \in \mathbb{C} : |\varphi(t)| = r\}, \quad r > 1 \tag{10.22}$$

is thus an ellipse, the image under ψ of the circle $\{|z| = r\}$. It is an easy exercise to check that E_r has foci $\{-1, 1\}$ and semi axes $\frac{1}{2}(r + r^{-1})$ and $\frac{1}{2}(r - r^{-1})$. Then, in view of (10.18) and (10.19) the series

$$\sum a_n P_n^{\alpha,\beta}(t) \tag{10.23}$$

hence converges absolutely (and uniformly) for t in the (open) domain D_R enclosed by the ellipse E_R.

Indeed, for each $r, 1 < r < R$, we have absolute convergence for $t \in E_r$ and hence, since (10.23) is a series of polynomials, convergence is uniform on $\overline{D_r}$. Thus, if (10.18) holds, (10.23) converges uniformly on compact subsets of D_R and defines there an analytic function. Nehari [**145**] investigated (in the case of Legendre polynomials P_n, where $\alpha = \beta = 0$ as in Szegő's original theorem) the analytic continuation of this analytic function and proved the following.

THEOREM 10.4. *Let $\{a_n\}$ be a sequence of complex numbers satisfying (10.18) and f the analytic function defined by*

$$f(t) := \sum_{n=0}^{\infty} a_n P_n(t) \tag{10.24}$$

for t in the open set D_R bounded by the ellipse with foci $\{-1, 1\}$, the sum of whose semi-axes is R. Then, f is analytically continuable across the boundary point t_0 of ∂D_R if and only if the analytic function

$$g(s) := \sum_{n=0}^{\infty} a_n s^n, \quad |s| < R \tag{10.25}$$

is analytically continuable across the point s_0 of its circle of convergence, where $s_0 = \varphi(t_0)$ (or, what is equivalent, $t_0 = \frac{1}{2}(s_0 + s_0^{-1})$).

To illustrate how remarkable Nehari's theorem is, let us note some immediate consequences:

COROLLARY 10.1. *Under the hypothesis of Theorem 10.4 if the function $f(t)$ given by (10.24) is analytically continuable across $t_0 \in \partial D_R$, the same is true for each of the series:*

$$\sum_{n=0}^{\infty} a_n P_{n+1}(t), \tag{10.26}$$

$$\sum_{n=0}^{\infty} a_{n+1} P_n(t), \tag{10.27}$$

$$\sum_{n=0}^{\infty} n a_n P_n(t). \tag{10.28}$$

PROOF. If (10.24) is continuable across t_0 then (10.25) is continuable across s_0 (by Theorem 10.4), and hence so is $\sum_{n=0}^{\infty} a_n s^{n+1}$. Using the theorem in the opposite direction, we conclude that (10.26) is continuable across t_0. The proofs for (10.27) and (10.28) are similar and we leave them as an (easy) exercise. □

Let us restate Nehari's theorem with a change of variables that would make it amenable to our methods and also put it into a context where the Legendre polynomials, being just a member in the vast hierarchy of Jacobi polynomials, play no specific role, thus allowing a far reaching generalization.

If we substitute $t = \cos w$ in (10.19) we get

$$\lim_{n\to\infty} |P_n^{\alpha,\beta}(\cos w)|^{1/n} = e^{|Imw|}, \tag{10.29}$$

for $w \in \mathbb{C} \setminus \mathbb{R}$. Consequently, if (10.18) holds the series

$$\sum_{n=0}^{\infty} a_n P_n^{\alpha,\beta}(\cos w). \tag{10.30}$$

converges absolutely and uniformly in the strip

$$S_\rho := \{|Im\{w\}| < \rho\}, \quad \rho = \log R > 0. \tag{10.31}$$

Nehari's theorem then says that (10.30) for $\alpha = \beta = 0$ is analytically continuable across the boundary point w_0 of this strip if and only if (10.25) is continuable across s_0 where $\frac{1}{2}(s_0 + s_0^{-1}) = \cos w_0$. Now, (10.25) is continuable across s_0 if and only if the same is true for the Laurent series

$$\sum_{n=0}^{\infty} a_n (s^n + s^{-n}), \tag{10.32}$$

or what comes to the same, setting $s = e^z$, if and only if

$$\sum_{n=0}^{\infty} a_n \cos nz, \tag{10.33}$$

which converges absolutely for $z \in S_\rho$ is analytically continuable across w_0. Finally, as in the context of Szegő's theorem, $\cos nz = T_n(\cos z)$ where $T_n = P_n^{-1/2,-1/2}$ denotes the Chebyshev polynomial of the first kind, and we can now reformulate Nehari's theorem as follows.

THEOREM 10.5. *If $\{a_n\}$ are complex numbers satisfying (10.18), then the two series (10.30) for $\alpha = \beta = 0$, and*

$$\sum_{n=0}^{\infty} a_n \cos nw = \sum_{n=0}^{\infty} a_n T_n(\cos w), \tag{10.34}$$

each of which converges and defines an analytic function in the strip S_ρ defined by (10.31), are analytically continuable across precisely the same boundary points of this strip.

Thus, we can interpret Nehari's theorem so that exchanging Legendre polynomials in (10.30) for Chebyshev polynomials does not alter the location of singularities on the boundary of the strip of convergence. This, of course, as above suggests a plausible generalization that we could replace Legendre polynomials in Theorem 10.5 with $\{P_n^{\alpha,\beta}\}$ for any choice $\alpha, \beta > -1$. This more general assertion was indeed proved by Gilbert [**64**] using the integral transform quite similar to Nehari's original proof (also, cf. [**199**], [**202**]). In terms of the original variables, Gilbert's theorem is this:

THEOREM 10.6 (Gilbert). *Theorem* 10.4 *holds if in place of* $P_n(t)$ *we write* $P_n^{\alpha,\beta}(t)$ *for any fixed choice of* $\alpha, \beta > -1$. *In other words, all the functions (which have the same ellipse of convergence) are analytically continuable across precisely the same boundary points of the ellipse.*

Before sketching the PDE proof of this result based on ideas similar to our proof of Szegő's theorem but quite different in several pivotal instances, let us once again stress an important point regarding normalization of Jacobi polynomials. When we say, e.g., that the Chebyshev polynomial T_n "is" a Jacobi polynomial we are referring to the formula

$$P_n^{(-1/2,-1/2)} = b_n T_n,$$

where

$$b_n = \frac{1 \cdot 3 \cdots (2n-1)}{2 \cdot 4 \cdots 2n}.$$

In formulating Nehari type theorems, it makes no difference whether we use T_n or $P_n^{(-1/2,-1/2)}$, because the sequence $\{b_n\}$ has the following property: the Taylor series $\sum_{n=0}^{\infty} b_n s^n$ and $\sum_{n=0}^{\infty} b_n^{-1} s^n$ each have radius of convergence 1, and each is analytically continuable across each point $s \neq 1$ of the unit circle. This follows, e.g., because each of the series is a hypergeometric series $F(a, b, c; s)$ for suitable values a, b, c (see [**187**, formula (4.21)]) and hence satisfies the corresponding Hadamard's multiplication of singularities theorem (see [**21**, **33**]). Thus, the series

$$\sum_{n=0}^{\infty} a_n s^n, \quad \sum_{n=0}^{\infty} a_n b_n s^n, \text{ and } \sum_{n=0}^{\infty} a_n b_n^{-1} s^n$$

all have precisely the same singularities on $\{|s| = 1\}$. Other "nuisance" factors in place of b_n, e.g., $P_n^{\alpha,\beta}(1) = \binom{n+\alpha}{n}$, that may arise through differences in normalization can be dealt with similarly. They will never affect the location of singularities, and we will tacitly ignore them from now on.

Now let us sketch a PDE proof of the Nehari-Gilbert theorem mostly concentrating on several differences compared to the arguments already presented in the proof of Szegő's theorem from the previous section. The details omitted here can be found in the original paper [**41**].

Let us summarize in the following assertion the result of adapting to the current context the calculations leading to Lemma 10.1 in the previous section. The detailed derivation is left as an exercise - cf. [**41**].

LEMMA 10.2. *Suppose* $\{a_n\}$ *satisfy* (10.18). *Then the series*

$$u = u_{\alpha,\beta}(z, w) = \sum_{n=0}^{\infty} a_n P_n^{\alpha,\beta}(\cos w) \cos(n+\sigma) z, \tag{10.35}$$

where $\sigma := \frac{\alpha+\beta+1}{2}$ *is holomorphic in the domain* Ω_ρ, $\rho = \log R$,

$$\Omega_\rho := \{z \in \mathbb{C}^2 : |Imz| + |Imw| < \rho\}, \tag{10.36}$$

(in view of (10.29)*) and satisfies the PDE*

$$Mu = 0, \tag{10.37}$$

where

$$M := \frac{\partial^2 u}{\partial w^2} - \frac{\partial^2 u}{\partial z^2} + \frac{(\alpha - \beta) + (\alpha + \beta + 1)\cos w}{\sin w} \frac{\partial u}{\partial w} - \left(\frac{\alpha + \beta + 1}{2}\right)^2 u.$$

Note that M in (10.37) is slightly different from its Szegő's theorem counterpart since instead of e^{nz} factors we are using $\cos(n+\sigma)z$. This accounts for the appearance of the last term in (10.37). However, the main properties of M are almost the same:

(i) The principal part of M is $\frac{\partial^2 u}{\partial w^2} - \frac{\partial^2 u}{\partial z^2}$.

(ii) M is linear, and its coefficients are meromorphic in $\mathbb{C}^2$ and have simple polar singularities only on the set $\{\sin w = 0\}$.

Recall that the Nehari-Gilbert theorem will be proved if we can show:

CLAIM. If $u_{\alpha,\beta}(0,w)$ (which is holomorphic in the strip $S_\rho := \{|Imw| < \rho\}$ is analytically continuable across the boundary point ζ, then $\sum_{n=0}^{\infty} a_n \cos nw$ is continuable across ζ and vice versa.

It is easy to see that $\sum_{n=0}^{\infty} a_n \cos nw$ and $\sum_{n=0}^{\infty} a_n \cos(n+\sigma)w$, both of which converge in S_ρ, are continuable across precisely the same boundary points of S_ρ. Hence, the Claim can be reformulated as the following, equivalent, assertion.

PROPOSITION 10.1. *The function $u_{\alpha,\beta}(0,w)$ is analytically continuable across $w = \zeta$ (where $|Im\zeta| = \rho$) if and only if $u_{\alpha,\beta}(z,0)$ is continuable across $z = \zeta$.*

Indeed, the continuability of each of the functions

$$\sum_{n=0}^{\infty} a_n \cos nw,$$

$$\sum_{n=0}^{\infty} a_n \cos(n+\sigma)w,$$

and

$$\sum_{n=0}^{\infty} a_n P_n^{\alpha,\beta}(\cos w)\cos(n+\sigma)z$$

across ζ are equivalent (in the last case, because the coefficients $P_n^{\alpha,\beta}(1) = \binom{n+\alpha}{n}$ satisfy the remark following Theorem 10.6).

To see the proof of Proposition 10.1, let us carry out our more preliminary reduction, embodied in the next two lemmas, where the set up is the same as in Proposition 10.1.

LEMMA 10.3. *If $u_{\alpha,\beta}(0,w)$ is analytically continuable across $w = \zeta$, with $|Im\zeta| = \rho$, then $u_{\alpha,\beta}(z,w)$ is analytically continuable to a neighborhood of $(0,\zeta)$ in $\mathbb{C}^2$.*

LEMMA 10.4. *If $u_{\alpha,\beta}(z,0)$ is analytically continuable across $z = \zeta$, with $|Im\zeta| = \rho$, then $u_{\alpha,\beta}(z,w)$ is analytically continuable to a neighborhood of $(\zeta,0)$ in $\mathbb{C}^2$.*

PROOF OF LEMMA 10.3. Let g denote an analytic function on a neighborhood U of ζ in the complex plane such that $g(w) = u_{\alpha,\beta}(0,w)$ for $w \in U \cap S_\rho$. Consider now the *holomorphic* Cauchy problem of finding v holomorphic in a neighborhood of $(0,\zeta)$ in $\mathbb{C}^2$, and satisfying

$$Mv = 0, \tag{10.38}$$

$$v(0,w) = g(w), \tag{10.39}$$

$$\frac{\partial v}{\partial z}(0,w) = 0. \tag{10.40}$$

This is a non-characteristic problem that has a unique solution v holomorphic on some neighborhood V of $(0, \zeta)$ in $\mathbb{C}^2$. If $w_0 \in U \cap S_\rho$ is sufficiently near to ζ then on a neighborhood of $(0, w_0)$ in $\mathbb{C}^2$, $u_{\alpha,\beta}$ satisfies $Mu_{\alpha,\beta} = 0$ and $u_{\alpha,\beta}$ has the same Cauchy data as v, i.e., (10.38) - (10.40) hold. Hence $u_{\alpha,\beta} = v$ on this neighborhood and v provides the analytic continuation of $u_{\alpha,\beta}$ across $(0, \zeta)$ as was to be proved. □

To prove Lemma 10.4 we must invoke the Mapel-Hemmati (Guerra) extension [**140**] of the Cauchy-Kovalevskaya Theorem mentioned in the previous section since the lower order terms in the equation have a simple polar singularity on on $\{w = 0\}$. The rest of the argument is the same, and we shall omit it. Thus, our Proposition 10.1 can be reformulated as the following proposition that would imply the Nehari-Gilbert Theorem.

PROPOSITION 10.2. *The function $u_{\alpha,\beta}(z, w)$ is analytically continuable across the boundary point $(\zeta, 0)$ of Ω_ρ if and only if it is continuable across $(0, \zeta)$.*

The proof will only use the fact that $u_{\alpha,\beta}$ is holomorphic in Ω_ρ, satisfies $Mu = 0$ and is *even* with respect to each of the variables z, w (this is why we needed $\cos nz$ versus e^{nz} in Szegő's theorem). First, we single out a simple corollary from the Bony-Schapira theorem on propagation of real-analyticity for hyperbolic analytic equations - cf. [**41**].

LEMMA 10.5. *Let $u(x_1, x_2)$ be real-analytic in a neighborhood of the closed interval $[0, 1]$ of the x_1-axis and satisfy there*

$$\frac{\partial^2 u}{\partial x_1^2} - \frac{\partial^2 u}{\partial x_2^2} + A_1 \frac{\partial u}{\partial x_1} + A_2 \frac{\partial u}{\partial x_2} + A_3 u = 0, \tag{10.41}$$

where the A_i are real-analytic in a neighborhood of the closed triangle $\overline{T}$ with vertices $(0, 0)$, $(1, 0)$, and $(1/2, 1/2)$. Then u extends real-analytically to a neighborhood of $\overline{T}$.

The proof rests on the Bony-Schapira construction of a globalizing family, and we shall omit it. The details can be found in [**41**, Lemma 3.6]. In our case the coefficients in (10.41) are meromorphic, yet the poles are confined to a very simple variety $\{\sin w = 0\}$. Then, invoking Theorem 10.3 used in the previous part of this chapter, we extend Lemma 10.5 to our situation. Here is the precise statement (cf. [**41**, Lemma 3.7]).

LEMMA 10.6. *Let u be as in Lemma 10.5 except that the coefficients A_i in (10.41) are assumed to be meromorphic in a complex neighborhood $V \subset \mathbb{C}^2$ of the closed triangle $\overline{T}$. Let Ω_t be the globalizing family for Lemma 10.5 and assume further that the singularity set of the coefficients A_i in V consists of a finite number of complex analytic curves that meet the boundaries $\partial\Omega_t$ transversally. Then, u still extends real-analytically to a neighborhood of $\overline{T}$.*

REMARK 10.5. In our proof of Proposition 10.2 we shall only need to assume that the singularities of the coefficients are confined to the complex line $\{z_2 = 1/2\}$. (The routine check that the globalizing family Ω_t constructed for Lemma 10.5 satisfies the hypothesis of Lemma 10.6 for our specific case of singularities confined to complex lines "parallel" to the z_1-axis can be found in [**41**].)

Now Proposition 10.2 will follow from the following slightly more general result. (We keep the same notation as before and write u for $u_{\alpha,\beta}$ with α, β fixed.)

PROPOSITION 10.3. *Suppose for some $\rho > 0$ that u is holomorphic in the domain $\Omega_\rho \subset \mathbb{C}^2$ defined in (10.36) and satisfies the equation*

$$\frac{\partial^2 u}{\partial w^2} - \frac{\partial^2 u}{\partial z^2} + B_1 \frac{\partial u}{\partial w} + B_2 \frac{\partial u}{\partial z} + B_3 u = 0, \tag{10.42}$$

where $B_i := B_i(z, w) = \frac{C_i(z,w)}{\sin w}$ and C_i are holomorphic on a neighborhood of $\overline{\Omega_\rho}$ $(i = 1, 2, 3)$. Then for $\zeta \in \mathbb{C}$ with $|Im\zeta| = \rho$ we have:

(a) *If u extends holomorphically across $(\zeta, 0)$ and $(-\zeta, 0)$ it extends across $(0, \zeta)$ and $(0, -\zeta)$.*

(b) *If u extends holomorphically across $(0, \zeta)$ and $(0, -\zeta)$ it extends across $(\zeta, 0)$ and $(-\zeta, 0)$.*

For the proof of (a) one merely needs via a complex linear transformation "move" the triangle with vertices $(-\zeta, 0)$, $(\zeta, 0)$, and $(0, \pm\zeta)$ to the triangle T in Lemma 10.5, and then apply the lemma. (The singularities of the coefficients do not matter since the target points stay away from the singularities. The hypothesis in (a) already takes care of these singularities.) For the proof of (b) we need the "heavier" hand of Lemma 10.6 since the coefficients of the PDE (10.42) have singularities at the "target points" $(\pm\zeta, 0)$. (The rather tedious details and the construction of the desired globalizing family can be found in [**41**, Section 3]. We omit them simply in order not to burden the reader with too many technicalities, opting for a general overview of the method.)

REMARK 10.6. Nehari also gives a result connecting the global analytic continuation of u and f. As in the previous section dealing with Szegő's theorem (see Theorem 10.1) our methods above also provide such assertions (cf. Remark 10.4 above concerning the Mittag-Leffler star). Nevertheless, the Nehari method based on explicit integral transforms has, whenever it is applicable, great power and precision and is by no means superseded by the PDE methods. Yet, the methods we have outlined here seem promising to apply to rather general eigenfunction expansions in any number of variables. By presenting it here we hope to prompt further research in this direction.

Notes

Szegő proved his theorem in [**186**] using ingeniously integral representations of axially symmetric harmonic functions. Shortly thereafter Nehari [**145**] published his theorem. The latter was followed by several substantial generalizations from Gilbert [**64**] and his school and later on by Zayed and his co-authors [**199, 201, 202**].

Further results extending Nehari's results to Faber and Appel polynomial expansions can be found in [**142**] and in the refernces cited there. (The latter results do not seem amenable to the PDE methods we emphasize here.) These later results all use the ideas from Szegő and Nehari's work based on far reaching generalizations of Hadamard's argument in his celebrated multiplication of singularites theorem. The PDE "high ground" to Szegő's and Nehari's theorem is from [**41, 42**]. The extended Cauchy-Kovalevskaya theorem for characteristic Cauchy problem of a special kind that we used in the proofs is from Mapel-Hemmati's paper [**140**]. The PDE methods look promising and one hopes, based on these ideas, to further extend

Szegő's and Nehari's theorems to expansions of eigenfunctions of other constant coefficients differential operators.

CHAPTER 11

The Reflection Principle

Let us try to resolve Question C stated in Chapter 1. First, we recall the setting for the classical Schwarz reflection principle in the plane. From now on by a non-singular analytic Jordan arc γ we will understand an object satisfying either one of the following (equivalent) definitions:

(i) $\gamma = \{z \in \mathbb{C} : z = f(t), 0 \le t \le 1, f$ is analytic and one-to-one in some neighborhood of $[0,1]\}$.

(ii) $\gamma = \{(x,y) \in \mathbb{R}^2 : \varphi(x,y) = 0$, where φ is a restriction of a holomorphic function $\varphi(X,Y)$ in a $\mathbb{C}^2$-neighborhood of γ such that $\nabla\varphi|_\gamma \neq (0,0)\}$.

1. The Schwarz function of a curve

PROPOSITION 11.1 (The Schwarz function of a curve). *Let γ be an analytic Jordan arc. Then there exists a (unique) function $S(z)$ analytic in a neighborhood of γ such that $\bar{z} = S(z)$ on γ.*

The function $S(z)$ is called the *Schwarz function* of γ.

EXAMPLE 11.1. If $\gamma = \mathbb{R}$, then $S(z) = z$.

EXAMPLE 11.2. If $\gamma = \{z : |z - z_0| = R\}$, then

$$S(z) = \overline{z_0} + \frac{R^2}{z - z_0}.$$

PROOF. Indeed, using the definition (ii) and the change of variables $z = X + iY$, $w = X - iY$ $(= \bar{z}$ in $\mathbb{R}^2)$, we obtain the equation of γ in the form $\Phi(z, \bar{z}) = 0$, where Φ is a holomorphic function of two complex variables. Note that

$$\left.\frac{\partial\Phi}{\partial\bar{z}}\right|_\gamma = \frac{1}{2}\left.\left(\frac{\partial\Phi}{\partial x} + i\frac{\partial\Phi}{\partial y}\right)\right|_\gamma \neq 0.$$

By the implicit function theorem we can solve that equation for $\bar{z}$ and obtain $\bar{z} = S(z)$, with the (implicit) function S being analytic in a neighborhood of γ. □

REMARK 11.1. If γ is closed, then $S(z)$ is analytic in a ring-like domain containing γ and since $S(z) = \bar{z}$ on γ, S has a single-valued branch near γ.

The following example illustrates this remark.

EXAMPLE 11.3. Let $\gamma = \left\{(x,y) \in \mathbb{R}^2; \frac{x^2}{a^2} + \frac{y^2}{b^2} = 1, a^2 - b^2 = 1\right\}$ be an ellipse. Solving a quadratic equation we easily find

$$S(z) = \left(a^2 + b^2\right) z - 2ab\left(z^2 - 1\right)^{\frac{1}{2}}.$$

Note that solving the quadratic equation one obtains two possible candidates for $S(z)$, yet only one of them equals $\bar{z}$ on Γ. (We leave checking this latter assertion as an easy exercise for the reader.)

The following theorem is one of the many ways to formulate the celebrated Schwarz Reflection Principle.

THEOREM 11.1. *Let γ be an analytic arc and $z^0 \in \gamma$. There exists a neighborhood V of z^0 homeomorphic to a disk and a bijective map $R_\gamma = R : V \to V$ such that:*

(i) *R is anti-conformal.*
(ii) *$R\,|_\gamma =$ identity.*
(iii) *If V_1 and V_2 are the components of $V \setminus \gamma$, $R(V_1) = V_2$.*
(iv) *R is an involution, i.e., $R \circ R =$ identity.*

PROOF. If we parametrize γ in terms of the arclength s as in $z = z(s)$, we note that on γ we have $S(z(s)) = \overline{z(s)}$ and

$$\frac{\overline{dz}}{ds} = \frac{dS(z)}{ds} = S'(z)\,\frac{dz}{ds},$$

so $S'(z(s)) = \overline{T(s)}^{2}$, where $T(s) = \frac{dz}{ds}$ is the unit tangent vector. In particular, $|S'(z)| \equiv 1$ on γ, so $S(z)$ is univalent near z^0. Now set

$$R(z) := \overline{S(z)}.$$

(i) is obvious and (ii) holds by the definition. $(R \circ R)(z) = R\left(\overline{S(z)}\right) = \overline{S(\overline{S(z)})}$ is analytic and $R \circ R = \overline{S(z)} = z$ on γ. So $R \circ R \equiv z$. Hence, it remains to show (iii). By Taylor's theorem we can write

$$S(z) = S\left(z^0\right) + S'\left(z^0\right)\left(z - z^0\right) + o\left(\left|z - z^0\right|\right),$$

for $\left|z - z^0\right|$ small, so that

$$R(z) = \overline{S\left(z^0\right)} + \overline{S'\left(z^0\right)}\left(\bar{z} - \overline{z^0}\right) + o\left(\left|z - z^0\right|\right).$$

Set

$$R^{\#}(z) = z^0 + T^2\left(\bar{z} - \overline{z^0}\right).$$

(Without loss of generality we can assume that $z^0 = 0$ and $T = (1, 0)$.) Note then that $R^{\#}$ is simply the reflection about the tangent line to γ at z^0 and (iii) now follows. □

COROLLARY 11.1 (Schwarz reflection principle).

(i) *Let $D \subset \mathbb{C}$ be a domain and $\sigma \subset \partial D$ be an analytic arc. Suppose that f is analytic in D and continuous in $\overline{D}$, and let $f(D) = E$ be the image of D under f so that $f(\sigma) = \tau \subset \partial E$ is an analytic arc. Then f extends analytically across σ and for points $z \in D$ sufficiently near σ the following "reflection law" holds*

$$f\left(R_\sigma(z)\right) = R_\tau(f(z)), \tag{11.1}$$

where R_σ and R_τ are the reflection maps with respect to the arcs σ and τ, respectively.

(ii) *Let u be harmonic in D, continuous in $\overline{D}$ and vanishing on σ. Then u extends as a harmonic function across σ and for $z \in D$ sufficiently close to σ the following "reflection law" (RL) holds*

$$u(z) + u\left(R_\sigma(z)\right) \equiv 0. \tag{11.2}$$

PROOF. (i) Shrinking D if necessary without loss of generality we can assume that $R_\sigma(D)$ is on another side of σ. This is guaranteed by (iii) of Theorem 11.1. Define $\hat{f}$ in $R_\sigma(D)$ by

$$\hat{f}(z) = R_\tau\left(f\left(R_\sigma^{-1}(z)\right)\right) = R_\tau \circ f \circ R_\sigma(z).$$

$\hat{f}$ is analytic in $R_\sigma(D)$ and continuous in $\overline{R_\sigma(D)}$. Moreover, $\hat{f}\big|_\sigma = f|_\sigma$. Hence, by Morera's theorem $\hat{f}$ is an extension of f across σ.

(ii) First of all, applying a conformal mapping if necessary we can assume that $\sigma \subset \mathbb{R}$. Define $\tilde{u}(x,-y) = -u(x,y)$. Then $\tilde{u}$ is harmonic in $R_\sigma(D)$ and smooth and continuous in $\overline{R_\sigma(D)}$. Moreover,

$$\tilde{u}|_\sigma = u|_\sigma = 0$$

and

$$\left.\frac{\partial \tilde{u}}{\partial y}\right|_{(x,0)} = \lim_{y\to 0^+} \frac{\tilde{u}(x,0) - \tilde{u}(x,-y)}{0-(-y)} = \lim_{y\to 0^+} \frac{u(x,y)}{y} = \left.\frac{\partial u}{\partial y}\right|_{(x,0)}.$$

By the Kovalevskaya theorem (Theorem 6.1) $\tilde{u}$ is an extension of u and (ii) follows. □

2. E. Study's interpretation of the Schwarz reflection principle

Consider the "real" wave equation

$$\frac{\partial^2 u}{\partial x \partial y} = 0.$$

Let γ be a smooth curve, say a C^2-curve, given by $\{y = s(x)\}$, where s is a C^2, strictly monotone decreasing function. Let $u = f(x) + g(y)$ be a C^2-solution of the wave equation in the region bounded by γ and the lines $x = a$, $y = b$, continuous up to the boundary, and such that $u|_\gamma = 0$. Let $a' = s^{-1}(b)$, $b' = s(a)$, $A = \left(s^{-1}(b), b\right)$, $B = (a, s(a))$. Then u extends as a C^2 solution to $[a,a'] \times [b,b']$ and the following reflection principle holds:

$$u(P) + u(Q) = 0,$$

where $P = (a,b)$ and $Q = \left(s^{-1}(b), s(a)\right)$. We call such P and Q *symmetric* with respect to the curve γ.

Indeed,

$$\begin{aligned} u(P) &= f(P) + g(P) \\ &= f(B) + g(A) \\ &= -g(B) - f(A) \text{ (since } u|_\gamma = 0) \\ &= -g(Q) - f(Q) \\ &= -u(Q). \end{aligned}$$

Turning back to the Laplace equation in $\mathbb{R}^2$, let γ be an analytic arc in $\mathbb{R}^2$, u satisfy $\Delta u = 0$ in a neighborhood of γ and $u|_\gamma = 0$. Then u extends as a holomorphic function to a $\mathbb{C}^2$-neighborhood of γ. Also, as we know, γ is the intersection of the holomorphic hypersurface $\Gamma = \left\{(X,Y) \in \mathbb{C}^2 : \varphi(X,Y) = 0\right\}$ with $\mathbb{R}^2$.

By changing variables $z = X + iY$, $w = X - iY$, we can write the equation of Γ in the form $w = S(z)$, where S is the Schwarz function of γ and if we are

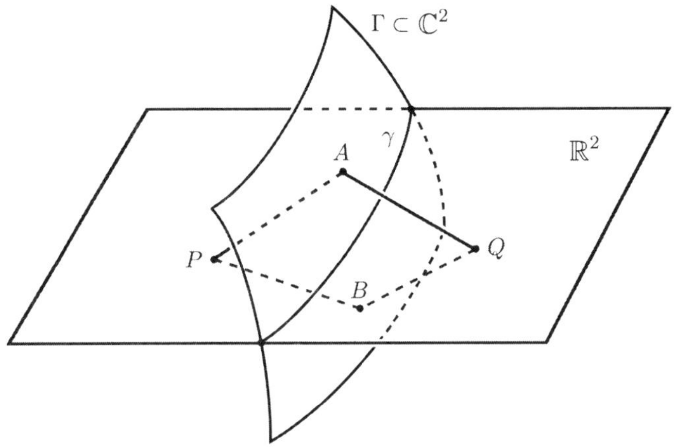

FIGURE 1. Study's interpretation of the Schwarz reflection principle.

sufficiently near γ, we can assume that $S(z)$ is univalent (cf. the proof of Theorem 11.1). Then, u satisfies

$$4\frac{\partial^2 u}{\partial z \partial w} = 0$$

near Γ and, since u is holomorphic, $u|_\Gamma = 0$. Now, $u = f(z) + g(w)$, where f and g are analytic functions of one variable. Then Study's interpretation of the reflection principle is as follows (see Fig. 1). Take $P = \left(z^0, \overline{z^0}\right) \in \mathbb{R}^2$ near γ and pass a complex line $M = \left\{(z,w) : z = z^0\right\}$ through P. Then $M \cap \Gamma = B = \left(z^0, S\left(z^0\right)\right)$. Similarly, the complex line $N = \left\{(z,w) : w = \overline{z^0}\right\}$ meets Γ at a point $A = \left(S^{-1}\left(\overline{z^0}\right), \overline{z^0}\right)$. Pass through A and B the complex lines M' and N', respectively:

$$M' = \left\{(z,w) : z = S^{-1}\left(\overline{z^0}\right)\right\},$$
$$N' = \left\{(z,w) : w = S\left(z^0\right)\right\}.$$

Set $Q := M' \cap N' = \left(S^{-1}\left(\overline{z^0}\right), S\left(z^0\right)\right)$. However, $Q \in \mathbb{R}^2$!. Indeed, $S^{-1}\left(\overline{z^0}\right) = \overline{S\left(z^0\right)}$ since according to Theorem 11.1, $z^0 = \overline{S\left(\overline{S\left(z^0\right)}\right)}$ and hence $\overline{z^0} = S\left(\overline{S\left(z^0\right)}\right)$. The rest of the argument extends word-for-word and (11.1) follows with $Q = R_\tau(P)$.

3. Failure of the reflection law for other operators

H. Lewy [**131**] showed that if u satisfies a second order PDE with the Laplacian in the principal part in a domain D adjacent to the real axis and $u|_{\mathbb{R}} = 0$, then u extends to the mirror image D' of D with respect to $\mathbb{R}$. Even if u satisfies a "linear relation" on $\mathbb{R}$, e.g., $u + u_y = 0$ on $\mathbb{R}$, it still extends to D'. He has also given a counterexample showing that the latter property fails in $\mathbb{R}^3$. However in regard to point-to-point reflection laws like (11.1) the situation for other operators even slightly varying from the Laplacian (or, the wave operator) is drastically different.

Let γ be an analytic curve in $\mathbb{R}^2$ and u satisfy the "Helmholtz" equation

$$\frac{\partial^2 u}{\partial x \partial y} + \lambda^2 u = 0 \tag{11.3}$$

near γ, $\lambda > 0$, and $u|_\gamma = 0$.

THEOREM 11.2. *If for two points P and Q sufficiently close to γ there exists a constant $k := k(P, Q)$ such that the "reflection law"*

$$u(P) + ku(Q) = 0, \tag{11.4}$$

holds for all u satisfying (11.3) near γ and vanishing on γ, then γ must be a straight line.

We will need a lemma that goes back to B. Riemann. Recall that if

$$\mathcal{L} = \frac{\partial^2}{\partial x \partial y} + a(x, y) \frac{\partial}{\partial x} + b(x, y) \frac{\partial}{\partial y} + c(x, y)$$

is a hyperbolic differential operator with a, b, c entire functions of two variables its adjoint is defined by

$$\mathcal{L}^{\#} u = \frac{\partial^2 u}{\partial x \partial y} - \frac{\partial}{\partial x}(au) - \frac{\partial}{\partial y}(bu) + cu.$$

The Riemann function $R_{\mathcal{L}} := R(x, y; \xi, \eta)$ at a point (ξ, η) for the operator $\mathcal{L}$ is defined as the solution of the following Cauchy-Goursat problem

$$\begin{cases} \mathcal{L}^{\#} R = 0 \text{ near } (\xi, \eta), \\ R(\xi, y; \xi, \eta) = \exp\left(\int_\eta^y a(\xi, \tau) d\tau\right), \\ R(x, \eta; \xi, \eta) = \exp\left(\int_\xi^x b(\tau, \eta) d\tau\right). \end{cases} \tag{11.5}$$

Note that from (11.5) it follows that $r(y) := R(\xi, y; \xi, \eta)$ satisfies $r_y - ar = 0$ on $\{x = \xi\}$, while $r(x) := R(x, \eta; \xi, \eta)$ satisfies $r_x - br = 0$ on $\{y = \eta\}$. Furthermore, observe that $R(\xi, \eta; \xi, \eta) = 1$ and as we noted in Chapter 5, $R(x, y; \xi, \eta)$ is an entire function. Moreover, since it is not hard to verify the symmetry property of the Riemann function namely, $R_{\mathcal{L}}(x, y; \xi, \eta) = R_{\mathcal{L}^{\#}}(\xi, \eta; x, y)$, it follows that $R(x, y; \xi, \eta)$ is in fact an entire function of all four variables.

LEMMA 11.1 (Riemann's lemma). *Let $\gamma = \{(x, y) : y = s(x)\}$ be a real-analytic curve that is non-characteristic with respect to $\mathcal{L}$. So, in particular, s and s^{-1} are real-analytic near γ. Let u be a solution of $\mathcal{L}u = 0$ near γ. For all points $P\left(x^0, y^0\right)$ sufficiently close to γ we have*

$$u(P) = \frac{1}{2}(uR)|_A + \frac{1}{2}(uR)|_B - \int_A^B (U dy - V dx),$$

where

$$R = R\left(x, y; x^0, y^0\right), \quad A = \left(s^{-1}\left(y^0\right), y^0\right), \quad B = \left(x^0, s\left(x^0\right)\right),$$

and

$$U = aRu + \frac{1}{2} Ru_y - \frac{1}{2} R_y u,$$
$$V = bRu + \frac{1}{2} Ru_x - \frac{1}{2} R_x u.$$

PROOF. The proof is a rather straightforward calculation. Let us separate the following assertion whose proof is left as an exercise.

ASSERTION.

$$0 = R(\mathcal{L}u) - u\left(\mathcal{L}^{\#}R\right) = U_x + V_y.$$

Assuming the assertion let G denote the region bounded by PA, PB and the arc AB. We have (applying Green's formula):

$$\begin{aligned} 0 &= \iint_G \left(R\mathcal{L}u - u\mathcal{L}^{\#}R\right) \\ &= \oint_{\partial G} U\,dy - V\,dx \qquad (11.6)\\ &= \int_A^B (U\,dy - V\,dx) - \int_P^B U\,dy - \int_P^A V\,dx. \end{aligned}$$

Now, using the properties of the Riemann function we compute

$$\begin{aligned} -\int_P^B U\,dy &= -\int_P^B \left[(aR - R_y)\,u + \frac{1}{2}\,(Ru)_y\right] dy = -\frac{1}{2}\,(Ru)|_P^B \\ &= \frac{1}{2}\,R(P;P)\,u(P) - \frac{1}{2}\,R(B;P)\,u(B) \\ &= \frac{1}{2}\,u(P) - \frac{1}{2}\,(Ru)|_B\,, \end{aligned}$$

and similarly,

$$-\int_P^A V\,dx = -\int_P^A \left[(bR - R_x)\,u + \frac{1}{2}\,(Ru)_x\right] dx = \frac{1}{2}\,u(P) - \frac{1}{2}\,(Ru)|_A\,. \qquad (11.7)$$

So, combining (11.6)–(11.7),

$$u(P) = \frac{1}{2}\,(uR)|_A + \frac{1}{2}\,(uR)|_B - \int_A^B (U\,dy - V\,dx). \qquad (11.8)$$

□

For the proof of Theorem 11.2 we shall need the following Corollary of Riemann's Lemma:

COROLLARY 11.2. *In particular, if* $u = 0$ *on* γ,

$$u(P) = \frac{1}{2}\int_A^B R\left(\frac{\partial u}{\partial x}\,dx - \frac{\partial u}{\partial y}\,dy\right). \qquad (11.9)$$

PROOF OF THEOREM 11.2. The operator $\mathcal{L}$ is now $\mathcal{L} := \frac{\partial^2}{\partial x \partial y} + \lambda^2$. We assume that P and Q are sufficiently close to

$$\begin{cases} \mathcal{L}u = 0, \\ \partial^\alpha u = \partial^\alpha[(y - s(x))g(x,y)], \quad |\alpha| \le 1, \end{cases} \qquad (11.10)$$

with g being a polynomial, are real-analytic at P and Q. Recall (cf. Example 5.1 in Chapter 5) that for our operator $\mathcal{L}$

$$R\left(x, y; x^0, y^0\right) = J_0\left(\lambda\sqrt{\left(x - x^0\right)\left(y - y^0\right)}\right),$$

where J_0 is the zero Bessel function. Replacing u on γ by the Cauchy data in (11.10), we obtain

$$\left(\frac{\partial u}{\partial x}\,dx - \frac{\partial u}{\partial y}\,dy\right)\Big|_\gamma = -2gs'(x)dx\Big|_\gamma .$$

So, in view of (11.9) we have

$$u(P) = \int_{A_P}^{B_P} J_0\left(\lambda\sqrt{(x-x^0)(y-y^0)}\right) g(x,y)s'(x)\,dx,$$

where we put $A = A_P = \left(s^{-1}\left(y^0\right), y^0\right)$, $B = B_P\left(x^0, s\left(x_0\right)\right)$ to stress their dependence on P. First, we show that (11.3) implies that P and Q must be "symmetric" with respect to γ. Suppose they are not. Without loss of generality we can assume then that A_Q is inside A_PB_P while B_P is inside A_QB_Q (other possible configurations can be treated similarly). By the Weierstrass approximation theorem [**162**, Ch. 7] we find polynomials g so that $|g(x,y)|$ is arbitrarily small on (A_Q, B_Q) while the integral $\int_{A_P}^{A_Q} J_0\left(\lambda\sqrt{(x-x^0)(y-y^0)}\right) g(x,y)s'(x)\,dx$ is larger than some fixed positive number η. This is possible since $s' \neq 0$ and $J_0 > 0$ near the origin. We see then, using (11.10) for P and Q, respectively, that $|u(Q)|$ can be made arbitrarily small, say $\leq \epsilon$, while $|u(P)| \geq \eta - \epsilon$. Hence, $u(P)$ cannot equal $-ku(Q)$. Thus, P and Q must be symmetric with respect to γ, i.e., $Q = \left(s^{-1}\left(y^0\right), s\left(x^0\right)\right)$.

Denote by

$$\mathcal{K}(x,y) := J_0\left(\lambda\sqrt{(x-x_0)(y-y_0)}\right) + kJ_0\left(\lambda\sqrt{(x-s^{-1}(y^0))(y-s(x^0))}\right).$$

The Reflection Law (11.4) implies that for all polynomials g we have $(A = A_P = A_Q, B = B_P = B_Q)$:

$$\int_A^B \mathcal{K}(x,y)gs'(x)\,dx = 0.$$

Therefore,

$$J_0\left(\lambda\sqrt{(x-x^0)(y-y^0)}\right) + kJ_0\left(\lambda\sqrt{(x-s^{-1}(y^0))(y-s(x^0))}\right) \equiv 0 \tag{11.11}$$

on γ. In particular, taking $(x,y) = \left(x^0, s\left(x^0\right)\right)$ and using the fact that $J_0(0) = 1$, we conclude that $k = -1$. Hence

$$J_0\left(\lambda\sqrt{(x-x^0)(y-y^0)}\right) = J_0\left(\lambda\sqrt{(x-s^{-1}(y^0))(y-s(x^0))}\right)$$

on γ and since J_0 is monotone decreasing on the positive semi-axis near the origin we must have $\left(x-x^0\right)\left(y-y^0\right) = \left(x-s^{-1}\left(y^0\right)\right)\left(y-s\left(x^0\right)\right)$ on γ. Therefore, γ must be a line. □

By using Study's change of variables $z = X+iY$, $w = X-iY$, that reduces the "real" Helmholtz's operator $\Delta + \lambda^2$ to the complex hyperbolic operator $4\frac{\partial^2}{\partial z \partial w} + \lambda^2$ and applying a similar argument we obtain the following.

COROLLARY 11.3. *Let γ be an analytic curve on $\mathbb{R}^2$. If for all solutions u of $\Delta u + \lambda^2 u = 0$ vanishing on γ the Reflection Law*

$$u(P) + ku(Q) = 0 \tag{RL}$$

holds for just two points P and Q sufficiently close to γ, then $k = -1$ and γ must be a line passing through the midpoint of PQ and orthogonal to PQ.

REMARK 11.2. For the operator $\mathcal{L} = \frac{\partial^2}{\partial x \partial y}$ the Riemann function equals 1 identically. Hence the representation formula (11.8) becomes

$$u(P) = \frac{1}{2} \int_A^B \left(\frac{\partial u}{\partial x}\, dx - \frac{\partial u}{\partial y}\, dy \right).$$

Therefore arguing as in the proof of the theorem we can show that if (RL)

$$u(P) + ku(Q) = 0$$

holds for all u satisfying $\mathcal{L}u = 0$, $u|_\gamma = 0$, then P and Q must be symmetric with respect to γ and $k = -1$.

Similarly, for $\mathcal{L} = \Delta$, the Laplace operator, the Reflection Law (i.e., the Schwarzian Reflection Law) holds near γ if and only if P and Q are symmetric with respect to γ, i.e., $Q = R(P)$, or equivalently, $K_P \cap \Gamma = K_Q \cap \Gamma$ where $P = (x_P, y_P)$, $Q = (x_Q, y_Q)$, and

$$K_P = \left\{ (X, Y) \in \mathbb{C}^2 : (X - x_P)^2 + (Y - y_P)^2 = 0 \right\}$$

and

$$K_Q = \left\{ (X, Y) \in \mathbb{C}^2 : (X - x_Q)^2 + (Y - y_Q)^2 = 0 \right\}$$

are isotropic cones in $\mathbb{C}^2$ emanating from P and Q, respectively, and $\Gamma = \{(X, Y) : X - iY = S(X + iY)\}$ is the "complexified" curve γ while S is, as usual, the Schwarz function of γ.

We leave as a straightforward exercise for the reader to check that P, Q being symmetric with respect to γ indeed is equivalent to the isotropic cones (in $\mathbb{C}^2$ just pair of complex lines!) emanating from P, Q respectively meet on the complexification Γ of γ at two points. As we shall see in the following chapter, the geometry in higher dimensions becomes much more complicated: the isotropic cones being irreducible hypersurfaces that intersect Γ along a surface of complex codimension two. Thus, the condition that they meet on Γ along precisely the same variety (the Study condition) is much more exclusive presenting a major obstacle to point-to-point reflection in higher dimensions.

COROLLARY 11.4. *Let $\Gamma = \{(x_1, x_2, x_3) : (x_1, x_2, 0) \in \gamma\}$ be a cylinder in $\mathbb{R}^2$ with base γ, an analytic curve. If the Reflection Law*

$$u(P) + ku(Q) = 0$$

holds for all harmonic functions u vanishing on Γ for P and Q sufficiently close to Γ, then Γ is a plane and P, Q are symmetric with respect to Γ.

PROOF. If $x_3^P \neq x_3^Q$, take $u(x_1, x_2, x_3) = \left(x_3 - x_3^P\right) v(x_1, x_2)$ where v is harmonic near γ, vanishes on γ and $v\left(x_1^0, x_2^0\right) \neq 0$. Then u is harmonic, vanishes on γ, $u(P) = 0$, and $u(Q) \neq 0$. So, we can assume $x_3^P = x_3^Q = 0$. By the previous remark, P and Q must also be symmetric with respect to γ. For all $\lambda > 0$, consider functions

$$u(x_1, x_2, x_3) = u^0(x_1, x_2)\, e^{\lambda x_3},$$

so that $u^0(x_1, x_2)\big|_\gamma = 0$ and satisfy $\Delta_{(x_1, x_2)} u^0 + \lambda^2 u^0 = 0$. Then all such u's are harmonic and vanish on Γ. Then, the Reflection Law implies that the hypothesis of Corollary 11.3 is satisfied for all u^0, and therefore γ must be a line. ☐

Notes

The term "Schwarz Function" was introduced by Ph. Davis in his book [**31**]. In presenting the Schwarz reflection principle via Theorem 11.1 we follow H. Shapiro's book [**176**], although Corollary 11.1 is very close to the reasoning in [**31**]. E. Study's beautiful interpretation of the reflection principle via the wave equation is from his paper [**184**], though he reasons right away for the "complex case". Apparently, his paper fell into oblivion and the idea was rediscovered again in the 1950's by a number of mathematicians, most notably by H. Lewy [**131**] and P. Garabedian [**56**]. In their papers, they also associated the Schwarzian Reflection with that for solutions of real wave equations. Theorem 11.2 or, more precisely, its complex analogue Corollary 11.3, is implicitly contained in [**111**] as an Ansatz in proving Corollary 11.4. Riemann's Lemma can be found, e.g., in Hadamard's classic [**72**]. Detailed discussions pertinent to Remark 11.2 can be found in [**111**], and in a more expanded form, in [**176**]. A far-reaching extension of Corollary 11.4 is contained in [**40**] (also see the following chapter). Finally, a "positive result" going the opposite direction from Corollary 11.2 has been found in [**169**]. There, the value at a point P of a solution of the Helmholtz equation vanishing on a curve γ is calculated by means of certain integrals along paths joining the point Q, symmetric to P, to γ.

In [**38**] Ebenfelt, as a corollary of his analytic continuation results, studied the surfaces in $\mathbb{R}^n$, with $n \geq 3$ for which a point-to-point reflection principle for the Helmholtz equation does hold. In [**4**] the authors applied the techniques in [**169**] to obtain nonlocal results for the reflection principle for polyharmonic functions of arbitrary order. A different approach was recently suggested in [**3**]. There has been revived activity on the subject recently. In [**15**] the authors extended the reflection principle in two dimensions for harmonic functions for which a linear combination of a function and its normal derivative vanish on the curve, extending the classical results of H. Lewy [**131**] for the line. This incidentally fails miserably in higher dimensions - [**131**]. For further results and references see [**167**]. Theorem 11.2 and Corollary 11.3 have been extended to allow not only point masses but distributions of finite order - [**2, 159**]. Further extensions to more general elliptic operators in the plane are contained in [**167, 168**]. Nevertheless the problem of point-to-compact set reflection, the so-called antenna problem, remains open in the plane even for Helmholtz equations. The reflection formulae in [**168, 169**] involve integrals over curves joining the reflected point with the "reflecting" curve γ. Thus, the intriguing question of whether one can "cut the umbilical cord" is still unresolved. Although Corollary 11.4 precludes point-to-point reflection across any cylinder in $\mathbb{R}^3$ except for the degenerate case of a plane, the harmonic functions vanishing on the surface of a spherical or elliptical cylinder can be extended globally; see [**59, 60**].

CHAPTER 12

The Reflection Principle (continued)

In the previous chapter we have studied the Reflection Law (RL) for two points P, Q in the plane and a real-analytic arc Γ. Namely, we were interested in the existence of a constant k only depending on P and Q such that

$$u(P) + ku(Q) = 0 \tag{RL}$$

holds for every function u harmonic near the arc Γ and vanishing on it.

In this chapter we shall discuss the possibility of extending the (RL) to higher dimensions. Recall now two cases for which the Reflection Principle does hold in all dimensions.

(1) Taking for a "reflecting" surface the hyperplane $\Gamma := \{x \in \mathbb{R}^n : x_n = 0\}$, we have
$$u(x', x_n) + u(x', -x_n) = 0,$$
for all u such that $\Delta u = 0$ near Γ and $u|_\Gamma = 0$.

(2) For the sphere $\Gamma = \left\{x \in \mathbb{R}^n : \sum_{i=1}^{n} x_i^2 = R^2\right\}$ we have,
$$u(x) + |x|^{2-n} u\left(R^2 \frac{x}{|x|^2}\right) = 0,$$
for all u harmonic near Γ such that $u|_\Gamma = 0$.

Let us prove the first statement. Define $u_1(x', x_n) = -u(x', -x_n)$ for $x_n < 0$. Then u_1 is harmonic in the half-space $\{x_n < 0\}$, continuous up to the boundary $\{x_n = 0\}$, and $u_1(x', 0) = 0$. Moreover, $\partial_{x_n} u_1(x', 0) = \partial_{x_n} u(x', 0)$. Hence, the harmonic function
$$U(x) = \begin{cases} u, & x_n \geq 0, \\ u_1, & x_n < 0. \end{cases}$$
has the same Cauchy data on $\{x_n = 0\}$ as does u, and, therefore, according to Holmgren's uniqueness theorem coincides with u in $\{x_n < 0\}$. This proves (1). The proof of the second statement is similar and is left as an exercise.

EXERCISE 12.1. Prove the reflection law for the sphere stated in (2). Show that, while the function $|x|^{2-n} u\left(R^2 \frac{x}{|x|^2}\right)$ is indeed harmonic if and only if u is, the function $u\left(R^2 \frac{x}{|x|^2}\right)$ is not, except when $n = 2$.

1. The Study relation

In both cases (1) and (2) above, the symmetric points P and Q satisfy the following relation (to which we will refer as the "Study Relation" based on the discussion in Section 2 of Chapter 11):

$$K_P \cap \widehat{\Gamma} = K_Q \cap \widehat{\Gamma}, \tag{SR}$$

where K_P and K_Q are, respectively, the isotropic cones emanating from P and Q, i.e.,

$$K_P = \left\{ z \in \mathbb{C}^n : \sum_{j=1}^{n} \left(z_j - x_j^P\right)^2 = 0 \right\},$$

$$K_Q = \left\{ z \in \mathbb{C}^n : \sum_{j=1}^{n} \left(z_j - x_j^Q\right)^2 = 0 \right\},$$

$P = \left(x_j^P\right)$, $Q = \left(x_j^Q\right)$, $j = 1, \dots, n$ and $\widehat{\Gamma}$ is the "complexification" of Γ (i.e., $\widehat{\Gamma} := \{z \in \mathbb{C}^n : z_n = 0\}$ in the first case and $\widehat{\Gamma} := \left\{ z \in \mathbb{C}^n : \sum_{i=1}^{n} z_j^2 = R^2 \right\}$ in the second).

Indeed, for the hyperplane $\widehat{\Gamma} := \{z_n = 0\}$, we have

$$K_P \cap \widehat{\Gamma} := \left\{ \sum_{j=1}^{n} (z_j - x_j^P)^2 = 0, z_n = 0 \right\} = K_Q \cap \widehat{\Gamma},$$

since $Q = ((x^P)', -x_n^P)$. We leave verification of the second case as the following exercise.

EXERCISE 12.2. Verify that the Study Relation (SR) holds in the case of the sphere.

Then, the first question one may pose here is the following: if the reflection law (RL) holds for P, Q near a real-analytic surface Γ in $\mathbb{R}^n$, $n \geq 3$, for all harmonic functions u vanishing on Γ (and defined in a fixed neighborhood of Γ containing P and Q), does it imply that the Study Relation (SR) holds? Or, in other words, is (SR) necessary for the (RL) to hold?

The answer is in the affirmative [**40**]. Here, however, we will only present a much simpler result [**111**]. Let us first introduce the following definition.

DEFINITION 12.1. We say that a harmonic function $u(x)$ has a *polar singularity* at a point x^0 if it is harmonic in a punctured neighborhood of x^0 and there exists $N > 0$ such that $\left|x - x^0\right|^N \cdot u(x)$ is bounded near x^0.

THEOREM 12.1. *Suppose that the Reflection Law (RL) holds for an enlarged class of test functions, namely, for harmonic functions $u(x)$ vanishing on Γ and admitting finitely many polar singularities near Γ. Then the Study Relation (SR) holds for the points P, Q.*

We shall need the following lemma.

LEMMA 12.1. *Let u be harmonic in the punctured ball $B\left(x^0, R\right) \setminus \left\{x^0\right\}$ and suppose that u has a polar singularity at x^0. Then u can be written in the form*

$$u(x) = \frac{Q(x)}{\left|x - x^0\right|^m} + v(x),$$

where m is a positive integer, $v(x)$ is harmonic in $B\left(x^0, R\right)$ and Q is a polynomial relatively prime with $\left|x - x^0\right|^2$.

PROOF OF LEMMA 12.1. By the growth condition on $u(x)$ near the singularity $x = x^0$, u can be extended as a distribution $u \in \mathcal{D}'\left(B\left(x^0, r\right)\right)$ to the whole ball $B\left(x^0, R\right)$[1]. Thus,

$$\operatorname{supp} \Delta u \subset \left\{x^0\right\}.$$

By the well-known theorem of L. Schwartz (see, e.g., [**161**, Theorem 6.25]), we can then write

$$\Delta u = R(\partial)\left(\delta_{\{x^0\}}\right),$$

where R is a polynomial (of degree $\leq N$) and $\delta_{\{x^0\}}$ is the point-mass at x^0. Also we have

$$\delta_{\{x^0\}} = \Delta\left\{\frac{c_n}{\left|x - x^0\right|^{n-2}}\right\}.$$

Thus,

$$\Delta\left(u - R(\partial)\left(\frac{c_n}{\left|x - x^0\right|^{n-2}}\right)\right) = 0$$

in the sense of distributions and according to Weyl's Lemma (see [**183**], also, see [**161**, Thm. 8.12 and its corollary] for the general version for arbitrary elliptic operators)

$$v := u - R(\partial)\frac{c_n}{\left|x - x^0\right|^{n-2}}.$$

must be harmonic in $B\left(x^0, R\right)$. This completes the proof of the lemma. □

PROOF OF THEOREM 12.1. First observe that in view of the assumed reflection, u must have a polar singularity at Q provided it has a polar singularity at P. Applying Lemma 12.1 we can write u in the form

$$u(x) = \frac{Q_1(x)}{\left|x - x_P\right|^{m_1}} + \frac{Q_2(x)}{\left|x - x_Q\right|^{m_2}} + v(x), \tag{12.1}$$

where $Q_1(x)$, $Q_2(x)$ are polynomials and $v(x)$ is harmonic. Let $\xi \in K_P \cap \widehat{\Gamma}$ and assume that $\xi \notin K_Q \cap \widehat{\Gamma}$. Let us denote by V a domain in $\mathbb{C}^n$ to which u can be extended holomorphically with the exception of the singularity set $K_P \cup K_Q$. Since $u \equiv 0$ on $\widehat{\Gamma} \setminus (K_P \cup K_Q)$ then if $z \to \xi$ in V with $z \in \widehat{\Gamma} \setminus (K_P \cup K_Q)$, while the left hand side in (12.1) remains zero, the right hand side becomes unbounded. Indeed, the first term tends to infinity, the second remains bounded since $\xi \notin K_Q \cap \widehat{\Gamma}$ and the third term is always bounded. Therefore we arrive at a contradiction and hence ξ must lie in $K_Q \cap \widehat{\Gamma}$. □

[1]For example one can define such an extension simply by

$$(u, \varphi)_{\mathcal{D}, \mathcal{D}'} = \int_{B(x^0, R)} u(x)\left(\varphi(x) - \sum_{|\alpha| \leq N-1} \frac{1}{\alpha!}\partial^\alpha \varphi\left(x^0\right)\left(x - x^0\right)^\alpha\right) dx$$

for every $\varphi \in C_0^\infty\left(B\left(x^0, R\right)\right)$, where N is the order of the "pole" at the origin.

2. Reflection in higher dimensions

The following result ([**111**]) shows that unlike in the case $n = 2$, there is no generic reflection in higher dimensions.

THEOREM 12.2. *Let $\Gamma \subset \mathbb{R}^n$ be a smooth algebraic surface. If*

$$\text{Volume}\{P \text{ near } \Gamma : \exists Q \text{ such that } P, Q \text{ satisfy } (SR)\} > 0,$$

then Γ must be either a sphere or a plane.

It follows from (1) and (2) above that spheres and planes satisfy this hypothesis; for planes Q is simply the symmetric image of P while for spheres Q is the inversion of P.

The proof of Theorem 12.2 although based on elementary algebro-geometric calculations is rather tedious, and we omit it here referring the interested reader to [**111**].

The following result [**40**] provides a complete answer as to the validity of the (RL) for odd-dimensional spaces.

THEOREM 12.3. *If the Reflection Law (RL) holds for two points P, Q sufficiently close to the real-analytic hypersurface $\Gamma \subset \mathbb{R}^{2n+1}$, then Γ must be either a plane or a sphere.*

REMARK 12.1. Let us, without diving into too many overwhelming technical details, quantify for the reader what do we mean by the rather loose expression "sufficiently close", here and in the sequel. First of all we restrict ourselves to a neighborhood U of Γ such that Γ is nonsingular in U, and hence by a biholomorphic change of coordinates can be made into a hyperplane. Moreover, the same holds for the part of the complexification $\widehat{\Gamma}$ of Γ contained in the domain $\widehat{U} \subset \mathbb{C}^n$, the harmonicity hull of U, i.e., the domain in $\mathbb{C}^n$ to which all functions harmonic in U extend holomorphically - cf. Ch. 9. Also, we are assuming that $\widehat{\Gamma}$ is noncharacteristic with respect to the Laplacian in $\widehat{U}$. Further, by shrinking U and hence $\widehat{U}$, we can assume, in view of Theorem 3.1, that *all* solutions of *all* CP for the Laplacian with the entire data posed on $\widehat{\Gamma}$ extend holomorphically to $\widehat{U}$. Finally, we might be forced to further restrict U ever so slightly (cf. [**40**, Sec. 1]) in order to carry through certain topological arguments (sketched below in an intuitive manner) regarding integral representations of solutions of the CP via integrating over cycles in $\widehat{U} \cup \widehat{\Gamma}$. For the full details we refer the reader to [**40**, Sec. 1]; also cf. pp. 161-162 ibid.

As for the (RL) in an even number of variables the situation is far more delicate. The following example already shows that the (RL) holds in a much greater variety of cases than in odd dimensions.

EXAMPLE 12.1. Consider in $\mathbb{R}^4$ an axially symmetric surface

$$\Gamma = \left\{(x_1, x_2, x_3, x_4) : f(x_1, \rho) = 0, \rho = \sqrt{x_2^2 + x_3^2 + x_4^2}\right\},$$

where f is, for instance, an even entire function and ρ stands for the distance to the x_1 axis. The meridian curve γ defined by the equation $f(x_1, \rho) = 0$ in the plane (x_1, ρ), i.e., $\gamma := \{(x_1, \rho) : f(x_1, \rho) = 0\}$, is then symmetric about the x_1-axis.

By making the change of variables:

$$\xi = x_1 + i\rho, \quad \xi^* = x_1 - i\rho,$$

and solving the equation $f(x_1, \rho) = 0$ for ξ^* we can write the "complexification" $\widehat{\Gamma}$ of Γ in terms of the complex variables ξ, ξ^* as $\widehat{\Gamma} := \{x \in \mathbb{C}^4 : \xi^* = S(\xi)\}$. The analytic function $S = S(\xi)$ is the Schwarz function of the curve γ (cf. Chapter 11).

First, observe the following.

LEMMA 12.2. *The Schwarz function S preserves the x_1-axis, in other words, if $P = (z, 0, 0, 0)$ then $S(P) = (Z, 0, 0, 0)$.*

PROOF. In view of the definition of S and since γ is symmetric about the x_1-axis, i.e., $\bar{\xi} \in \gamma$ provided that $\xi \in \gamma$, it follows that

$$\bar{\xi} = S(\xi) \text{ and } \xi = S\left(\bar{\xi}\right) \text{ on } \gamma. \tag{12.2}$$

From (12.2) we have $\overline{S(\xi)} = S\left(\bar{\xi}\right)$ on γ, or conjugating,

$$S(\xi) = \overline{S(\bar{\xi})} \text{ on } \gamma. \tag{12.3}$$

Since both functions in (12.3) are analytic and equal on γ they coincide thus giving $S(P) = \overline{S\left(\overline{P}\right)} = \overline{S(P)}$ for all P on the x_1-axis implying the lemma. □

Let us now state a positive result from [**101**] regarding the (RL) for the case of axially symmetric hypersurfaces in $\mathbb{R}^4$.

THEOREM 12.4. *For points $P = (a, 0, 0, 0)$ on the axis of symmetry, the following (RL) holds:*

$$u(P) + (-S'(P))\, u(S(P)) = 0 \tag{12.4}$$

for all u harmonic near Γ and vanishing on it.

Before giving the proof of the theorem let us see how examples (1) and (2) fit into the context.

EXAMPLE 12.2 (Planes). In this case γ coincides with the imaginary axis $\gamma = \{i\rho\}$ (here $\Gamma := \{x_1 = 0\}$). The equation for γ is $\xi^* = -\xi$ and so, $S(\xi) = -\xi$. If $P = (a, 0, 0, 0)$, $S(P) = (-a, 0, 0, 0)$, $S'(P) = -1$. Then (12.4) simply yields for this case the well-known reflection law

$$u(a, 0, 0, 0) + u(-a, 0, 0, 0) = 0.$$

EXAMPLE 12.3 (Spheres). In this case $\Gamma = \{x \in \mathbb{R}^4 : x_1^2 + \rho^2 = 1\}$. The equation for γ becomes $\xi^* \xi = 1$ and thus $S(\xi) = \frac{1}{\xi}$. If $P = (a, 0, 0, 0)$, $S(P) = \left(\frac{1}{a}, 0, 0, 0\right)$ and, since $S'(\xi) = -\frac{1}{\xi^2}$ the relation (12.4) now becomes the familiar Kelvin reflection in the sphere in $\mathbb{R}^4$

$$u(a, 0, 0, 0) + \frac{1}{a^2}\, u\left(\frac{1}{a}, 0, 0, 0\right) = 0.$$

Let us also check the Study Relation for points

$$P = (a, 0, 0, 0), \quad Q = S(P) = (S(a), 0, 0, 0)$$

on the x_1 axis in the last example. We have (as before $\xi = x_1 + i\rho$, $\xi^* = x_1 - i\rho$):

$$K_P = \left\{x \in \mathbb{C}^4 : (x_1 - a)^2 + \rho^2 = 0\right\} = \left\{x \in \mathbb{C}^4 : (\xi - a)(\xi^* - a) = 0\right\}$$

and

$$K_{S(P)} = \left\{x \in \mathbb{C}^4 : (\xi - S(a))(\xi^* - S(a)) = 0\right\}.$$

Thus if $(\xi, \xi^*) \in K_P \cap \widehat{\Gamma}$ then $\xi^* = S(\xi)$ and $\xi = a$ or $S(\xi) = a$. If $\xi = a$ then $S(\xi) = S(a)$ and $(\xi, \xi^*) \in K_{S(P)} \cap \widehat{\Gamma}$. If $S(\xi) = a$ then $\xi = S^{-1}(a)$. Since $a = S(S(a))$ and so $S^{-1}(a) = S(a)$, $\xi = S(a)$ and hence $(\xi, \xi^*) \in K_{S(P)} \cap \widehat{\Gamma}$. It just remains to note that all the steps are reversible.

PROOF OF THEOREM 12.4. Let u be harmonic in a neighborhood of Γ, $u|_\Gamma = 0$. Let us separate the following claim.

ASSERTION. The function u can be written in the form $u = u_0 + u_1$ where the functions u_0 and u_1 are both harmonic, vanish on Γ and $u_0(x) = u_0(x_1, \rho)$ is axially symmetric while $u_1(x_1, 0, 0, 0) = 0$.

Assuming the assertion, let us continue with the proof of the theorem. Since u_1 vanishes on the x_1-axis, it suffices to prove (12.4) for axially symmetric harmonic functions $u(x) = u(x_1, \rho)$. It is easy to check that an axially symmetric function $u(x_1, \rho)$ is harmonic in $\mathbb{R}^4$ iff

$$\frac{\partial^2 u}{\partial x_1^2} + \frac{\partial^2 u}{\partial \rho^2} + \frac{2}{\rho}\frac{\partial u}{\partial \rho} = 0. \tag{12.5}$$

(It is crucial here that the dimension of the space is 4! For other dimensions, say n, the axially symmetric harmonic functions satisfy the equation similar to (12.5) but the constant 2 is replaced by $n-2$, so the following trick fails.) Thus, if $u(x_1, \rho)$ is harmonic then $v(x_1, \rho) = \rho u(x_1, \rho)$ is harmonic as a function of the variables x_1, ρ in the meridian plane. Indeed, (12.5) can be written as

$$\rho \Delta_{(x_1,\rho)} + 2\nabla_{(x_1,\rho)} u \cdot \nabla_{(x_1,\rho)} \rho = 0,$$

or, in other words, $\Delta_{(x_1,\rho)}(\rho u) = 0$. But now $v(x_1, \rho)$ satisfies the Schwarz Reflection Principle with respect to γ, thus

$$v(\xi) + v\left(\overline{S(\xi)}\right) = 0$$

for all ξ sufficiently close to γ. In other words,

$$\rho u(x_1 + i\rho) + Im\overline{S(x_1 + i\rho)} u\left(\overline{S(x_1 + i\rho)}\right) = 0. \tag{12.6}$$

Differentiating (12.6) with respect to ρ we obtain

$$u(\xi) + i\rho\frac{\partial u}{\partial \rho} + \frac{\partial}{\partial \rho}\left(Im\overline{S(\xi)}\right) u\left(\overline{S(\xi)}\right) + Im\overline{S(\xi)}\frac{\partial u}{\partial \rho}\left(\overline{S(\xi)}\right) = 0. \tag{12.7}$$

Since by Lemma 12.2 $Im(S(\xi)) = 0$ when $\rho = 0$ the relation (12.4) follows from (12.7) and the fact that according to the Cauchy-Riemann equations and Lemma 12.2

$$\frac{\partial}{\partial \rho}\left(Im\overline{S(\xi)}\right)\Big|_{\rho=0} = -S'(x_1).$$

□

PROOF OF ASSERTION. Let us now proceed with the proof of the assertion. If $u(x_1, x_2, x_3, x_4)$ is the given harmonic function, define

$$u_0(x_1, \rho) = \frac{1}{4\pi\rho^2}\int_{S(x_1,\rho)} u(x_1, y)\, dS_{(x_1,\rho)}(y), \quad \rho \neq 0,$$

where $y = (y_2, y_3, y_4) \in \mathbb{R}^3$, $S(x_1, \rho) = \left\{(x_1, y) : y \in \mathbb{R}^3 : y_2^2 + y_3^2 + y_4^2 = \rho^2\right\}$ is the 2-dimensional sphere centered at $(x_1, 0, 0, 0)$ of radius ρ, dS is a Lebesgue measure on $S(x_1, \rho)$. It is clear then that $\lim_{\rho\to 0^+} u_0(x_1,\rho) = u(x_1,0,0,0)$ and it is also clear that $u_0|_\Gamma = 0$. Let us now show that u_0 is harmonic. As observed above we must check that u_0 satisfies

$$\frac{\partial^2 u_0}{\partial x_1^2} + \frac{\partial^2 u_0}{\partial \rho^2} + \frac{2}{\rho}\frac{\partial u_0}{\partial \rho} = 0.$$

If $S^2(0,1)$ denotes the unit sphere in $\mathbb{R}^3$ and dS_y stands for the Lebesgue measure on it, u_0 can be expressed as

$$u_0(x_1,\rho) = \frac{1}{4\pi}\int_{S^2(0,1)} u((x_1,0,0,0)+\rho(0,y))\, dS_y.$$

Hence,

$$\frac{\partial^2 u_0}{\partial x_1^2} = \frac{1}{4\pi}\int_{S^2(0,1)} \frac{\partial^2 u}{\partial x_1^2}((x_1,0,0,0)+\rho(0,y))\, dS_y. \tag{12.8}$$

On the other hand,

$$\frac{\partial u_0}{\partial \rho} = \frac{1}{4\pi}\int_{S^2(0,1)} \nabla_{x'} u((x_1,0,0,0)+\rho(0,y))\cdot y\, dS_y,$$

where $x' = (x_2,x_3,x_4)$. Taking into account that for each $y \in S^2(0,1)$ the vector $\vec{y}$ is the exterior unit normal ν with respect to the ball

$$B(0,\rho) = \left\{(x_1,y) : y \in \mathbb{R}^3 : \|y\| \le \rho\right\},$$

the divergence theorem implies:

$$\begin{aligned}\frac{\partial u_0}{\partial \rho} &= \frac{1}{4\pi\rho^2}\int_{S(x_1,\rho)} \frac{\partial u}{\partial \nu}\, dS_{(x_1,\rho)}(y)\\ &= \frac{1}{4\pi\rho^2}\int_{B(0,\rho)} \Delta_{x'} u(x_1,x')\, dx'.\end{aligned} \tag{12.9}$$

(Here $\Delta_{x'} := \sum_{j=2}^{4} \frac{\partial^2}{\partial x_j^2}$.) Differentiating again with respect to ρ we obtain using (12.9)

$$\begin{aligned}\frac{\partial^2 u_0}{\partial \rho^2} &= -\frac{2}{4\pi\rho^3}\int_{B(0,\rho)} \Delta_{x'} u(x_1,x')\, dx'\\ &\quad + \frac{1}{4\pi\rho^2}\frac{\partial}{\partial \rho}\int_{B(0,\rho)} \Delta_{x'} u(x_1,x')\, dx'\\ &= -\frac{2}{\rho}\frac{\partial u_0}{\partial \rho} + A.\end{aligned} \tag{12.10}$$

Now, since u is harmonic $-\frac{\partial^2 u}{\partial x_1^2} = \Delta_{x'} u$, we calculate using spherical coordinates and (12.8):

$$
\begin{aligned}
A &= \frac{1}{4\pi\rho^2}\frac{\partial}{\partial\rho}\left\{\int_0^\rho\left(\int_{|y|=1}\Delta_{x'}u\,(x_1, ry)\,dS_y\right)r^2dr\right\}\\
&= \frac{1}{4\pi}\int_{S^2(0,1)}\Delta_{x'}u\,(x_1,\rho y)\,dS_y\\
&= -\frac{1}{4\pi}\int_{S^2(0,1)}\frac{\partial^2 u}{\partial x_1^2}\left((x_1,0,0,0)+\rho(0,y)\right)dS_y\\
&= -\frac{\partial^2 u_0}{\partial x_1^2}.
\end{aligned}
\tag{12.11}
$$

From (12.10) and (12.11) we obtain

$$\frac{\partial^2 u_0}{\partial \rho^2} = -\frac{2}{\rho}\frac{\partial u_0}{\partial\rho} - \frac{\partial^2 u_0}{\partial x_1^2}$$

and the proof of harmonicity of u_0 is concluded. Therefore the assertion follows by writing $u(x) = u_0(x) + u_1(x)$ with $u_1(x) := u(x) - u_0(x)$. □

3. The even-dimensional case

For a point $x^0 \in \mathbb{R}^n$ set

$$g\left(z - x^0\right) := \left(z_1 - x_1^0\right)^2 + \cdots + \left(z_n - x_n^0\right)^2,$$

the defining function of the isotropic cone $K_{\{x^0\}}$ with vertex at x^0.

Let Γ be a real-analytic hypersurface in $\mathbb{R}^n$ and $\widehat{\Gamma}$ its extension to $\mathbb{C}^n$. Consider two points $x^0, x^1 \in \mathbb{R}^n \setminus \Gamma$ that satisfy the Study Relation (SR), i.e.,

$$K_{\{x^0\}} \cap \widehat{\Gamma} = K_{\{x^1\}} \cap \widehat{\Gamma}. \tag{SR}$$

Let us now introduce the following definition.

Definition 12.2. Let the dimension n be an even number. We say that the points $x^0, x^1 \in \mathbb{R}^n$ satisfy the Strong Study Relation (SSR for short) if x^0, x^1 satisfy (SR) and

$$g\left(\cdot - x^0\right)\big|_{\widehat{\Gamma}} = \left(\lambda^{\frac{1}{p}} + h(\cdot)\left(g\left(\cdot - x^1\right)\big|_{\widehat{\Gamma}}\right)^p\right) g\left(\cdot - x^1\right)\big|_{\widehat{\Gamma}}, \tag{SSR}$$

where $p = \frac{n-2}{2}$, λ is a constant, and h is some holomorphic function.

Note that for $n = 4$, $p = 1$ we have that $g(\cdot - x^0)$ matches $\lambda g(\cdot - x^1)$ up to first order on $\widehat{\Gamma} \cap K_{x_j}$, $j = 0, 1$, since the term $h(\cdot)\left(g(\cdot - x^1)\right)^2$ vanishes to second order on $\widehat{\Gamma}$. Hence, for $n = 4$ Definition 12.2 simply means that

$$dg(\cdot - x^0)\big|_{\widehat{\Gamma}} = \lambda dg(\cdot - x^1)\big|_{\widehat{\Gamma}}.$$

Let us give an equivalent geometric description of SSR points for $n = 4$. Fix two points x^0 and x^1 in $V \setminus \Gamma$ and define a to be half the distance between the points. (Recall that V is a neighborhood that is inside the "sufficiently small" neighborhood U described in Remark 12.1.) Without loss of generality, performing translation and rotation if necessary, we can assume that $x^0 = (0, 0, 0, a)$ and $x^1 = (0, 0, 0, -a)$; let us denote the coordinates by $s = (s_1, s_2, s_3, s_4)$, and as before $s' = (s_1, s_2, s_3)$.

PROPOSITION 12.1. *Fix $n = 4$ with x^0, x^1, and s as above. The two points x^0 and x^1 are SSR points with respect to Γ and with constant λ if and only if $\widehat{\Gamma}$ is tangent to the complex sphere*

$$\Sigma = \left\{ \langle s', s' \rangle + \left(s_4 + \frac{1}{c} \right)^2 + a^2 = \frac{1}{c^2} \right\},$$

where

$$c = \frac{1 - \lambda}{a + a\lambda}$$

along the set

$$\Lambda = \{s_4 = 0\} \cap \{\langle s', s' \rangle + a^2 = 0\}.$$

If $\lambda = 1$, then $c = 0$, and the surface Σ becomes the hyperplane $\{s_4 = 0\}$.

PROOF. First assume that x^0 and x^1 are SSR points with respect to Γ with constant λ. It is easy to check that

$$K_0 \cap K_1 = \Lambda,$$

where K_j is the isotropic cone with vertex at x^j, for $j = 0, 1$. Since the points x^0 and x^1 are SR points, for $j = 0, 1$ we must have $\widehat{\Gamma} \cap K_j$ contained in Λ, and since this set is irreducible we have equality. To fix the ideas, assume that $\widehat{\Gamma}$ can be written as $s_4 = \psi(s')$ near each point on Λ (the general case being similar). Also, let g_- denote the function $g(\cdot - x^0)$ and g_+ the function $g(\cdot - x^1)$. It follows that

$$\frac{1}{2} dg_{\pm} = \langle s', ds' \rangle + (s_4 \mp a) ds_4.$$

Taking the restriction to $\widehat{\Gamma}$ and using the coordinates s' there we obtain

$$\frac{1}{2} dg_{\pm}\big|_{\widehat{\Gamma}} = \langle s' + (s_4 \mp a) \nabla \psi, ds' \rangle.$$

From this equation and Definition 12.2 we see that if the points are SSR points then

$$d\psi = -c \langle s', ds' \rangle$$

on Λ, where

$$c = \frac{1 - \lambda}{a + a\lambda}.$$

This is equivalent to $\widehat{\Gamma}$ being tangent to Σ along Λ. This proves the "only if" part of the proposition. The second part can be deduced if we notice the following. First, observe that all the steps above are reversible. Then, an elementary but technically tedious fact, established in [**40**], [Lemma 1.3], that $\widehat{\Gamma} \cap K_j$ is irreducible in a fixed neighborhood containing x_0, x_1 allows us to conclude that $\widehat{\Gamma} \cap K_j$ must equal Λ if Σ is tangent to $\widehat{\Gamma}$ along Λ. □

Loosely speaking, Proposition 12.1 states that the condition for two points to be SSR points with respect to Γ means that $\widehat{\Gamma}$ has an infinitesimal axial symmetry near the set Λ about the line connecting the two points.

Similarly, in higher even dimensions the analogous proposition holds, i.e., Definition 12.2 is equivalent to $\widehat{\Gamma}$ meeting a certain complex sphere (or hyperplane) with center on the line through the points x^0 and x^1 with pth order tangency along the intersection between the isotropic cones K_{x^0} and K_{x^1}. Let us illustrate the SSR condition by a couple of examples.

EXAMPLE 12.4. Let us consider axially symmetric hypersurfaces in $\mathbb{R}^n$, with $n \geq 4$ even. By choosing coordinates appropriately, such a surface Γ can be expressed by an equation of the form $f(x_1, \rho) = 0$, where f is a real analytic function of two variables with nonvanishing gradient along its zero locus and where $\rho^2 = x_2^2 + \cdots + x_n^2$. Geometrically, we can think of Γ as being obtained by revolving a curve

$$\gamma = \{(x_1, \rho) \in \mathbb{R}^2 : f(x_1, \rho) = 0\},$$

symmetric with respect to the x_1-axis, around the x_1-axis in $\mathbb{R}^n$ (where $\mathbb{R}^2$ is considered as a two-plane in $\mathbb{R}^n$ such that the x_1-axis in $\mathbb{R}^2$ coincides with the x_1-axis in $\mathbb{R}^n$). We call γ the meridian curve and $\mathbb{R}^2$ with coordinates (x_1, ρ) the meridian plane. First, let us return to axially symmetric surfaces in $\mathbb{R}^4$ already considered in Theorem 12.4. If we identify x^0 with its corresponding point $(t, 0) \in \mathbb{R}^2$ (where $t = x_1^0$) then the point x^1 corresponds to $(S(t), 0) \in \mathbb{R}^2$, where $S(x_1 + i\rho)$ denotes the Schwarz function of the meridian curve γ, and the constant k in (RL) equals $-S'(t)$. Let us verify that x^0 and x^1 so defined satisfy the SSR conditions with respect to Γ in $\mathbb{R}^4$. Introducing coordinates $\zeta = z_1 + i\rho$, $\zeta^* = z_1 - i\rho$ in the complexified meridian plane (z_1, ρ), we can write the equations of the isotropic cones K_{x^0}, K_{x^1} as follows:

$$\begin{cases} g(\cdot - x^0) = (z_1 - t)^2 + \rho^2 = (\zeta - t)(\zeta^* - t) = 0, \\ g(\cdot - x^1) = (z_1 - S(t))^2 + \rho^2 = (\zeta - S(t))(\zeta^* - S(t)) = 0. \end{cases} \tag{12.12}$$

K_{x^0} and K_{x^1} meet on the plane $z_1 = (t + S(t))/2$, and the intersections of K_{x^j} with $\widehat{\Gamma}$ meet the complexified meridian plane (ζ, ζ^*) at the two points $(S^{-1}(t), t)$ and $(t, S(t))$; note that $\widehat{\gamma} \subset \mathbb{C}^2 := \{\zeta^* = S(\zeta)\}$. Hence, in view of (12.12) we have on $K_{x^0} \cap \widehat{\Gamma} = K_{x^1} \cap \widehat{\Gamma}$

$$\begin{cases} dg(\cdot - x^0)\big|_{\widehat{\Gamma}} = ((S(\zeta) - t) + (\zeta - t)S'(\zeta))\, d\zeta, \\ dg(\cdot - x^1)\big|_{\widehat{\Gamma}} = ((S(\zeta) - S(t)) + (\zeta - S(t))S'(\zeta))\, d\zeta. \end{cases} \tag{12.13}$$

Thus, the SSR condition in $\mathbb{R}^4$ (in its equivalent formulation in Definition 12.2 for x^0 and x^1 at $(t, S(t))$ and $(S^{-1}(t), t)$ is fulfilled with

$$\lambda = -\frac{1}{S'(t)}$$

(for the second point we must keep in mind that $S = S^{-1}$, which follows from the fact that γ is symmetric with respect to the x_1-axis), as claimed above, and we arrive at the statement in Theorem 12.4.

Recall that, geometrically, the SSR condition in $\mathbb{R}^4$ means that $\widehat{\Gamma}$ is tangent, at the appropriate points, to the sphere given in Proposition 12.1. In higher even dimensions, $n \geq 6$, the SSR condition is more restrictive (it calls for higher order of contact between $\widehat{\Gamma}$ and the sphere), and the corresponding statement for axially symmetric hypersurfaces is not true in general as the following examples illustrate.

(i) Let $\gamma \subset \mathbb{R}^2$ be the ellipse

$$x_1^2 + \frac{\rho^2}{a^2} = 1,$$

where $a^2 \neq 1$, and let Γ be the corresponding axially symmetric ellipsoid in $\mathbb{R}^n$, $n \geq 4$ even. Since $\widehat{\Gamma}$ is a quadric, it cannot have an order of contact greater than one with another quadric unless the two quadrics are the same. Hence, since $a^2 \neq 1$, no two points are SSR points with respect to Γ in $\mathbb{R}^n$, with $n \geq 6$ even.

(ii) Let $\gamma \subset \mathbb{R}^2$ be the curve

$$x_1 = c(\rho^2 + 1)^k,$$

where $k \geq 2$ is an integer and $c > 0$ is small, and let Γ be the corresponding surface of revolution in $\mathbb{R}^n$, with $n \geq 4$ even. We claim that the points $x^0 = (1, 0, ..., 0)$ and $x^1 = (-1, 0, ..., 0)$ are SSR points with respect to Γ and with constant $\lambda = 1$ if and only if $n \leq 2k$. To prove this, we shall verify that the complex hyperplane $\{z_1 = 0\}$ in $\mathbb{C}^n$ meets $\widehat{\Gamma}$ with order of contact $k - 1$ along the intersection with the isotropic cones K_{x^0} and K_{x^1}, in accordance with the geometric interpretation of the SSR condition. By axial symmetry, it suffices to check that $\widehat{\gamma}$ in the complex meridian plane $\mathbb{C}^2$ meets the two points $(0, i)$ and $(0, -i)$ (the points in $\mathbb{C}^2$ that generate the intersection in $\mathbb{C}^n$ by revolution) and, moreover, that $\widehat{\gamma}$ meets the hyperplane $\{z_1 = 0\}$ with order of contact $k - 1$ at $(0, i)$ and $(0, -i)$. If we denote the variables in $\mathbb{C}^2$ by (z_1, ρ) (as above, we allow ρ to be complex here) then $\widehat{\gamma}$ is given by

$$z_1 = c(\rho^2 + 1)^k.$$

Obviously, the two points $(0, i)$ and $(0, -i)$ are on $\widehat{\gamma}$ and the order of contact between $\widehat{\gamma}$ and $\{z_1 = 0\}$ at each of the two points is, as claimed, $k - 1$. This proves the claim above, since x^0 and x^1 are SSR points with respect to a real hypersurface Γ and with constant $\lambda = 1$ if and only if $(n - 2)/2 \leq k - 1$.

The complete solution of the point-to-point reflection problem in even dimensions was given in the following theorem in [**40**].

THEOREM 12.5. *Let the dimension n be even. The (RL)*

$$u\left(x^0\right) + ku\left(x^1\right) = 0$$

holds, for all functions u harmonic near Γ and vanishing on Γ, if and only if x^0 and x^1 satisfy the (SSR) with $\lambda = \frac{1}{k}$.

The proof of Theorem 12.5 is rather technical and relies on several deep results in several complex variables beyond the scope of this book. Accordingly we limit our discussion to the four-dimensional case. To help the reader understand the nature of the SSR condition and why it is necessary and sufficient, let us discuss in detail the case $n = 4$ under a restricted hypothesis that reduces problem to the more concrete setting of the classical Huygens' principle for the wave equation.

Suppose that our surface $\widehat{\Gamma} = \{z \in \mathbb{C}^4 : f(z) = 0\}$ appears as a real hypersurface in the real subspace

$$W = i\mathbb{R}^3 \times \mathbb{R} = \{(ix', x_4) : x_j \in \mathbb{R}\}$$

(generically, the intersection of $\widetilde{\Gamma}$ with this subspace has real codimension two). Let $x^0 = (0, 0, 0, t)$ and $x^1 = (0, 0, 0, -t)$ be SR points. Obviously, we have $K_{x^0} \cap K_{x^1} \subset \{z_4 = 0\}$. Any u harmonic in U and vanishing on Γ extends as a holomorphic function into $\mathbb{C}^4$ and satisfies the wave equation

$$\sum_{j=1}^{3} \frac{\partial^2 u}{\partial x_j^2} - \frac{\partial^2 u}{\partial x_4^2} = 0$$

in W. In the setting of the wave equation, the cones $\tilde{I}_0 = K_{x^0} \cap W$ and $\tilde{I}_1 = K_{x^1} \cap W$ are referred to as "light cones" emanating from x^0 and x^1. By the hypothesis, they meet on $\widetilde{\Gamma}$ along the 2-dimensional sphere $S_t = \{(ix', 0) : |x'| = t\}$. Now, by

Kirchoff's formula (see e.g. [**93**, Ch. 5]), the value of u at x^0 can be calculated in terms of the Cauchy data on the plane $\{x_4 = 0\}$ along the sphere S_t only:

$$u(x^0) = \frac{C}{t^2} \int_{S_t} \left(t \frac{\partial u}{\partial x_4} + \sum_{j=1}^{3} \frac{\partial u}{\partial x_j} x_j \right) dS,$$

where C is some constant and dS denotes the Lebesgue measure on S_t taken with positive (with respect to the outward normal) orientation. Note that we have used the fact that u vanishes on S_t. Since $-\nabla g(\cdot - x^0) = 2(x', t)$ on the plane $\{x_4 = 0\}$ we can rewrite the above equation in the form

$$\begin{aligned} u(x^0) &= -\frac{C}{2t^2} \int_{S_t} \langle \nabla u, \nabla g(\cdot - x^0) \rangle dS \\ &= -\frac{C}{2t^2} \int_{S_t} \frac{\partial u}{\partial n} \langle \hat{n}, \nabla g(\cdot - x^0) \rangle dS, \end{aligned}$$

where $\hat{n}$ denotes the unit normal to $\tilde{\Gamma}$. Similarly, we have

$$u(x^1) = -\frac{C}{2t^2} \int_{S_t} \frac{\partial u}{\partial n} \langle \hat{n}, \nabla g(\cdot - x^1) \rangle dS.$$

Now, if (RL) holds, then in view of equations above and since $\partial u / \partial n$ runs over a dense set of the continuous functions on S_t as u runs over the space of functions harmonic in a fixed sufficiently small neighborhood U of $\tilde{\Gamma}$ and vanishing on $\tilde{\Gamma}$, it follows that

$$\langle \nabla g(\cdot - x^0) + k \nabla g(\cdot - x^1), \hat{n} \rangle = 0$$

on S_t. To fix ideas, assume that $\tilde{\Gamma}$ can be written as $\{x_4 = \phi(x')\}$. Then, since the intersection of $\tilde{\Gamma}$ with the plane $\{x_4 = 0\}$ contains the sphere S_t it follows that $\nabla \phi = \psi(x') x'$ on S_t. Combining this with the equation above and the definition of S_t, we find that ψ is constant on S_t,

$$\psi(x') = \frac{1}{t} \frac{1-k}{1+k}.$$

Noting that

$$\left\{ \begin{array}{l} -dg(\cdot - x^0)\big|_{\widehat{\Gamma}} = 2\langle x' + t\nabla\phi, dx' \rangle, \\ -dg(\cdot - x^1)\big|_{\widehat{\Gamma}} = 2\langle x' - t\nabla\phi, dx' \rangle. \end{array} \right. \tag{12.14}$$

we obtain that

$$dg(\cdot - x^0)\big|_{\tilde{\Gamma}} = \lambda dg(\cdot - x^1)\big|_{\tilde{\Gamma}},$$

where $\lambda = 1/k$ on S_t, and, therefore, everywhere on $K_{x^0} \cap \tilde{\Gamma} = K_{x^1} \cap \tilde{\Gamma}$. This is precisely the SSR condition in Definition 12.2.

As an example, for the sake of pedagogy and hopefully not being overly repetitive, let us analyze once again the (SSR) for the case of axially symmetric hypersurfaces Γ in $\mathbb{R}^4$ already studied in this chapter. As above x_1, ρ denote the meridian complex coordinates, $\xi = x_1 + i\rho$, $\xi^* = x_1 - i\rho$ and the complex extension $\widehat{\Gamma}$ of Γ to $\mathbb{C}^4$ is given by $\xi^* = S(\xi)$. As already seen in Theorem 12.4 the (RL) holds for all pairs of points $x^0 = (a, 0, 0, 0)$ and $x^1 = (S(a), 0, 0, 0)$. In this case:

$$g\left(\cdot - x^0\right) = \left(\xi - a\right)\left(\xi^* - a\right), \quad g\left(\cdot - x^1\right) = \left(\xi - S(a)\right)\left(\xi^* - S(a)\right),$$

and

$$g\left(\cdot - x^0\right)\big|_{\widehat{\Gamma}} = (\xi - a)(S(\xi) - a), \quad g\left(\cdot - x^1\right)\big|_{\widehat{\Gamma}} = (\xi - S(a))(S(\xi) - S(a)).$$

As before (cf. Example 12.2) one easily verifies the (SR):

$$\begin{aligned}K_{\{x^0\}} \cap \widehat{\Gamma} &= K_{\{x^1\}} \cap \widehat{\Gamma} \\ &= \left\{ x_1 = \frac{a+S(a)}{2} = \frac{a+S^{-1}(a)}{2}, \rho = \pm\frac{(a-S(a))}{2i} \right\} \\ &= \{(a, S(a))\} \cup \left\{ \left(S^{-1}(a), a\right) \right\}.\end{aligned}$$

We now have $n = 4$, so $p = 1$. Since $p = 1$, (SSR) simply says that

$$dg\left(\cdot - x^0\right)\big|_{\widehat{\Gamma}\cap K_{\{x^0\}}} = \lambda dg\left(\cdot - x^1\right)\big|_{\widehat{\Gamma}\cap K_{\{x^1\}}}.$$

We have

$$\begin{aligned}dg\left(\cdot - x^0\right)\big|_{\widehat{\Gamma}} &= \left((S(\xi) - a) + S'(\xi)(\xi - a)\right) d\xi, \\ dg\left(\cdot - x^1\right)\big|_{\widehat{\Gamma}} &= \left((S(\xi) - S(a) + S'(\xi)(\xi - S(a))\right) d\xi.\end{aligned}$$

Thus

$$\begin{aligned}dg\left(\cdot - x^0\right)_{(a,S(a))} &= (S(a) - a)\, d\xi, \\ dg\left(\cdot - x^1\right)_{(a,S(a))} &= -S'(a)(S(a) - a)\, d\xi,\end{aligned}$$

and (SSR)

$$dg\left(\cdot - x^0\right)_{(a,S(a))} = \lambda dg\left(\cdot - x^1\right)_{(a,S(a))}, \tag{12.15}$$

holds with $\lambda = -\frac{1}{S'(a)}$ (cf. Theorem 12.4). The same value for λ is obtained in (12.15) if we calculate both differentials at points on $\widehat{\Gamma} \cap K_{\{x^0\}}$, where $(\xi, \xi^*) = \left(S^{-1}(a), a\right)$, and recall (Lemma 12.2) that $S^{-1}(a) = S(a)$ and thus $S'(S(a)) = \frac{1}{S'(a)}$.

4. The odd-dimensional case

Finally, let us return to the discussion of the odd-dimensional case, Theorem 12.3. The proof requires techniques in several complex variables beyond the scope of this book (see [**40**] for full details). Here, we will merely explain some of the key ideas. Let us start by revisiting the two-dimensional case where we outline an approach based on arguments of Garabedian and Lewy [**56, 57, 131**] that had grown out of the ideas of Study [**184**] (also, cf. [**40**] for further discussions).

THEOREM 12.6. *Let $\Gamma \subset \mathbb{R}^2 \cong \mathbb{C}$ be a non-singular real-analytic curve in a neighborhood U of a point $z' = x' + iy'$. Then there exists a perhaps smaller neighborhood V of z' and an anti-conformal involution $R : V \to V$ such that $R\big|_\Gamma =$ id and*

$$u(z) + u(R(z)) = 0, \quad z \in V$$

for all u harmonic in V and vanishing on Γ.

PROOF. We can write the equation of Γ in the form

$$\overline{z} = S(z)$$

where S is the Schwarz function of Γ holomorphic in V, where $V \subset U$ is some neighborhood of z'. Shrinking V even further if necessary we can assume that S is also univalent in V. Imbed $\mathbb{R}^2$ into $\mathbb{C}^2$ as an anti-holomorphic plane $\{w = \overline{z}\}$, so

Γ is the intersection between the complex-analytic curve $\widehat{\Gamma} = \{w = S(z)\}$ and $\mathbb{R}^2$. Fix a point $z^0 \in V \setminus \Gamma$ and let u be harmonic in V and vanish on Γ. Then

$$u(z^0) = \frac{1}{2\pi}\int_{\gamma_\varepsilon}\left(u(\zeta)\frac{\partial}{\partial n_\zeta}\log|z^0-\zeta| - \frac{\partial u}{\partial n_\zeta}\log|z^0-\zeta|\right)ds_\zeta, \tag{12.16}$$

where $\gamma_\varepsilon = \{z \in \mathbb{C} : |z - z^0| = \varepsilon\}$ is a sufficiently small circle centered at z^0, n_ζ is the outer normal, and ds_ζ is the arclength element. As is known (see for example Corollary 8.2), u extends as a holomorphic function of z and w into a certain fixed $\mathbb{C}^2$-neighborhood $\hat{U}$ of $\widehat{\Gamma}$ which depends only on U and not on the function u. The form that we are integrating in (12.16) is a closed (multi-valued in $\mathbb{C}^2$) 1-form that has a logarithmic singularity on the isotropic cone K_0 with vertex at the point z^0,

$$K_{z^0} = \{(z - z^0)(w - \overline{z^0}) = 0\}.$$

Therefore, we can move the contour γ_ε homotopically (not homologically due to the multi-valuedness) within $\hat{U} \setminus K_{z^0}$ without changing the value of the integral (12.16). Our first goal is to deform γ_ε to a contour on $\widehat{\Gamma}$. For that purpose, note that γ_ε is homotopic in $\mathbb{C}^2 \setminus K_{z^0}$ to a contour γ_ε^1 that consists of two disjoint non-linked circles, each one surrounding the complex lines $\{z = z^0\}$ and $\{w = \overline{z^0}\}$ respectively, joined by a "handle". Indeed, $\mathbb{C}^2 \setminus K_{z^0}$ is homotopically equivalent to the three-sphere S^3 without two one-dimensional circles resulting from the intersection of K_{z^0} with S^3. Choosing a point on one of these circles and performing a stereographic projection with a pole at this point we obtain $\mathbb{R}^3$ with a line and a circle removed. The contour γ_ε becomes a loop tying them together. At this point it is easy to see that γ_ε is homotopic to the contour γ_ε^1 defined above. Having deformed γ_ε into γ_ε^1 we move γ_ε^1 continuously in $\mathbb{C}^2 \setminus K_{z^0}$ to a contour contained in $\widehat{\Gamma}$. Thus, γ_ε in (12.16) can be replaced by a closed contour γ_ε^2 on $\widehat{\Gamma}$ that consists of two small ε-circles σ_ε^1, σ_ε^2 surrounding the points $A = (S^{-1}(z^0), z^0)$ and $B = (z^0, S(z^0))$ and an arc σ_ε^3 joining σ_ε^1 and σ_ε^2 traveled twice. Since $u|_{\widehat{\Gamma}} = 0$, the equation (12.16) can be written in terms of an integral along the contour γ_ε^2:

$$u(z^0) = \frac{1}{4\pi i}\int_{\gamma_\varepsilon^2}\log(z - z^0)\log(w - \overline{z^0})\left(\frac{\partial u}{\partial z}dz - \frac{\partial u}{\partial w}dw\right). \tag{12.17}$$

Now observe that as $\varepsilon \to 0$ the integrals in (12.17) over the circles σ_ε^1, σ_ε^2 tend to zero, while that along σ_ε^3 tends to

$$u(z^0) = \frac{1}{2}\int_\gamma\left(\frac{\partial u}{\partial z}dz - \frac{\partial u}{\partial w}dw\right), \tag{12.18}$$

where $\gamma \subset \widehat{\Gamma}$ is an arc joining the points A and B (from A to B). The equation (12.18) follows from (12.17) since the logarithm in (12.17) on "one side of σ_ε^3" differs from that on "the other side" by $2\pi i$. Finally, applying the same argument to the point $z^1 = \overline{S(z^0)}$ we obtain

$$u(z^1) = \frac{1}{2}\int_{-\gamma}\left(\frac{\partial u}{\partial z}dz - \frac{\partial u}{\partial w}dw\right) = -u(z^0),$$

and the theorem follows with $R(z) = \overline{S(z)}$. □

Now, for $n = 3$ repeating the same argument while omitting certain salient topological details (see [**40**, pp. 158-166] for the details) we will end up with the integral over a two-dimensional topological disk D surrounded by $K_0 \cap \widehat{\Gamma}$ of a 2-form involving u and derivatives of the Newtonian kernel $N(\cdot - x^0) = \frac{1}{4\pi}|\cdot - x^0|^{-1}$ and terms involving derivatives of u times $N(\cdot - x^0)$. Now, $N(\cdot - x^0)$ changes sign as we go around the boundary C of the base since N has a square root, rather than logarithmic singularity on K_0. The orientation of D though changes sign after we go around C and end up on a second copy of D. Hence, instead of cancellation, we obtain

$$u(x^0) = 2\int_D \sum_{j=1}^{3}(-1)^j \frac{\partial u}{\partial z_j} N(z - x^0)\omega_j, \tag{12.19}$$

where

$$\omega_j = dz_1 \wedge \cdots \wedge \widehat{dz_j} \wedge \cdots \wedge dz_3,$$

with $\widehat{dz_j}$ indicating that dz_j is missing, and a similar formula to (12.19) holds for $u(x^1)$. Note that in view of the C-K theorem ∇u can run over the set of vector fields with arbitrary polynomials as coordinate functions, i.e., over a dense subset of continuous vector fields on the two-dimensional topological disk D. Thus, for the reflection law (RL) to hold we need to have

$$N(z - x^0) + kN(z - x^1) \equiv 0$$

on D, and hence, everywhere on $\widehat{\Gamma}$. Indeed, $\widehat{\Gamma}$ is a four real dimensional variety (two complex dimensional variety) in $\mathbb{C}^3$. Thus, ignoring routine changes of variables, what we are saying is simply that an analytic function of two complex variables in $\mathbb{C}^2$ is uniquely determined by its restriction to $\mathbb{R}^2$ - the restriction to the disk D uniquely identifies the Taylor coefficients. The latter is an equation of a sphere, or a plane. A similar but still more technical argument extends the theorem to all odd dimensions (see [**40**] for details).

Notes

The notion of Study Relation (SR) appeared in [**111**]. Theorem 12.2 has been proved in [**111**] (for $n = 3$). Theorems 12.3 and 12.5 that completely settled the problem of point-to-point reflection in higher dimensions were established in [**40**]. The notion of Strong Study Relation (SSR) has also been introduced in [**40**].

Theorem 12.4 appeared earlier in [**101**]. The idea of moving the cycle representing harmonic functions near a point x_0 via integrating an $(n-1)$-dimensional closed form that includes the fundamental solution (the Newtonian or logarithmic kernel) toward the complexified surface $\widehat{\Gamma}$ homotopically in $\mathbb{C}^n \setminus K_{x_0}$ goes back to Garabedian; cf. [**56**].

CHAPTER 13

Cauchy Problems and the Schwarz Potential Conjecture

Let $n = 2$ and $\mathcal{L} = \frac{\partial^2}{\partial z \partial w} + a(z, w)\frac{\partial}{\partial z} + b(z, w)\frac{\partial}{\partial w} + c(z, w)$ be a differential operator with all coefficients assumed to be entire functions.

Let Γ be a hypersurface in $\mathbb{C}^2$, non-characteristic with respect to $\mathcal{L}$, given by $\{w = S(z)\}$ or $\{z = S^{-1}(w)\}$ with S, S^{-1} holomorphic near Γ.

The following theorem was noticed by several authors, e.g., P. Garabedian, I.N. Vekua, P. Henrici, H. Lewy to name a few.

THEOREM 13.1. *The solution u of a Cauchy Problem*

$$\left\{ \begin{array}{l} \mathcal{L}u = 0, \\ \partial^\alpha (u - f) = 0, \ |\alpha| \leq 1 \ \text{on } \Gamma \end{array} \right. \tag{13.1}$$

with entire data f extends (perhaps as a multivalued function) along any path in $\mathbb{C}^2$ starting at a non-characteristic point on Γ provided that $S(z), S^{-1}(w)$ extend as holomorphic functions along that path.

PROOF. Take (z^0, w^0) sufficiently close to Γ, so the solution u is holomorphic at (z^0, w^0) by the C-K theorem. Represent $u(z^0, w^0)$ by Riemann's formula ((11.8), Lemma 11.1) applied in the context of $\mathbb{C}^2$. However, on Γ we have $w = S(z)$, $z = S^{-1}(w)$, $u = f(z, S(z)) = f(S^{-1}(w), w)$, $\frac{\partial u}{\partial w} = \frac{\partial f}{\partial w}(z, S(z)) = \frac{\partial f}{\partial w}(S^{-1}(w), w)$, $\frac{\partial u}{\partial z} = \frac{\partial f}{\partial z}(z, S(z)) = \frac{\partial f}{\partial z}(S^{-1}(w), w)$ and f along with the Riemann function R associated to $\mathcal{L}$ are entire (the latter in view of Corollary 5.3 and the remark preceding Lemma 11.1). Thus, the right hand side of (11.8) extends holomorphically to any point (z^0, w^0) in $\mathbb{C}^2$ as long as $S(z)$ and $S^{-1}(w)$ extend to that point. $\square$

Consider now a special (CP)

$$\left\{ \begin{array}{l} \dfrac{\partial^2 u}{\partial z \partial w} = 0, \\ \partial^\alpha (u - zw) = 0, \ |\alpha| \leq 1 \text{ on } \Gamma. \end{array} \right. \tag{13.2}$$

Denote the (unique) solution u_Γ. Then $u_\Gamma = F(z) + G(w)$, where F, G are analytic functions of one variable, and so

$$\left. \frac{\partial u_\Gamma}{\partial z} \right|_\Gamma = F'(z)|_\Gamma = w|_\Gamma = S(z)|_\Gamma .$$

Hence, $\frac{\partial u_\Gamma}{\partial z} = S(z)$. Similarly, $\frac{\partial u_\Gamma}{\partial w} = S^{-1}(w)$.

Therefore Theorem 13.1 can be restated to say that solutions to all Cauchy problems (13.1) with entire data extend holomorphically along any path starting at a non-characteristic point on Γ and avoiding all the singularities of the solution u_Γ of (13.2).

Restricting ourselves to $\mathbb{R}^2 = \{(z,w) : \bar{w} = z\}$ while changing variables to $z = x+iy$, $\bar{z} = x - iy$, and accordingly, the operator to $\mathcal{L} := \Delta + A\frac{\partial}{\partial x} + B\frac{\partial}{\partial y} + C$ (where A, B, C are entire functions), with

$$\gamma := \Gamma \cap \mathbb{R}^2 \subset \{(z,\bar{z}) : \bar{z} = S(z), S(z) \text{ is the Schwarz function of } \gamma\},$$

and recalling (cf. Chapter 11) that $S^{-1}(\bar{z}) = \overline{S(z)}$ we are led to the following corollary.

COROLLARY 13.1. *All solutions of (CP)*

$$\begin{cases} \mathcal{L}u = 0, \\ \partial^\alpha(u - f) = 0, \ |\alpha| \le 1 \text{ on } \gamma \end{cases} \tag{13.3}$$

with entire data f (as a function of two complex variables) extend along all paths in $\mathbb{R}^2$ provided that $S(z)$ is analytically continuable along those paths, or equivalently u_Γ $(= u_\gamma)$ defined by

$$\begin{cases} \Delta u_\Gamma = 0, \\ \partial^\alpha\left(u_\Gamma - |z|^2\right) = 0, \ |\alpha| \le 1 \text{ on } \gamma \end{cases} \tag{13.4}$$

is continuable along those paths.

REMARK 13.1 (Heuristic explanation of Leray's principle). Let us discuss the above result in depth while embarking on the path of trying to answer Question E from Sec. 5 in Ch. 1, thus previewing the ideas of Leray discussed in Chapter 19. To fix the ideas let us return to the real-variable setting of Lemma 11.1. Assume that our initial curve γ is algebraic, i.e.,

$$\gamma := \{\varphi(x,y) = 0\},$$

with φ a polynomial, so the implicitly defined function $y = s(x)$ and its inverse $x = s^{-1}(y)$ are globally-defined algebraic functions. Note that the characteristic (with respect to $\mathcal{L} = \frac{\partial^2}{\partial x \partial y}$ + lower order terms) points are precisely the points where either $s' = 0$ or ∞, i.e., the points where γ has either a horizontal or vertical tangent, i.e. $\{(x,y) : \varphi_x = 0 \text{ or } \varphi_y = 0\}$. The characteristic points are precisely the points where (locally), in view of the failure of the Cauchy-Kovalevskaya Theorem, solutions of the Cauchy Problem might become singular.

In fact, the solution of a very simple (CP)

$$\begin{cases} \frac{\partial^2}{\partial x \partial y} u_\gamma = 0, \\ u_\gamma \equiv xy \text{ on } \gamma, \end{cases} \tag{13.5}$$

already is singular at *all* characteristic points since as is easily seen $(u_\gamma)_x = s(x), (u_\gamma)_y = s^{-1}(y)$, and

$$u_\gamma = \int s(x)dx + \int s^{-1}(y)dy + \text{const}.$$

Furthermore, and this is crucial, u_γ is *only* singular on the lines $s(x) = y_0$, $s^{-1}(y) = x_0$ that are tangent to γ at characteristic points $(x_0, y_0) \in \gamma_{ch}$. For example, the curve

$$\gamma := \{(x,y) \in \mathbb{R}^2 : y = x(x-1)(x-2)\}$$

has no vertical tangents, since $s(x) = x(x-1)(x-2)$ has no singularities, while $s^{-1}(y)$ becomes singular along two horizontal lines $y = y_j$, $j = 1,2$, with

$$y_j = s(x_j)$$

where $x_1 = \frac{3+\sqrt{3}}{3}, x_2 = \frac{3+\sqrt{3}}{3}$ satisfy $s'(x_j) = 0$. Now, Riemann's formula (cf. (11.8), Lemma 11.1) and its consequence, Theorem 13.1, asserts that the solution of *any* Cauchy problem with entire data on γ might only become singular after continuing it from a non-characteristic piece of γ and hitting a point where s, s^{-1} are singular. In other words arriving at a point from which there exists a horizontal or vertical tangent to γ. Now, transplanting the latter argument to $\mathbb{C}^2$, we conclude that *all* possible singularities of the (CP) (13.2) are confined to a union Σ of complex lines $\{z = z_0\}$, $\{w = w_0\}$ in $\mathbb{C}^2$ tangent to the initial surface Γ. This Σ is precisely an example in a simple situation of $\mathbb{C}^2$, and constant coefficients in the principal part, of a developable hypersurface tangent to Γ at characteristic points. Thus, in a nutshell, Theorem 13.1 and Corollary 13.2 affirm that all possible singularities encountered by solutions of the (CP) (13.2) in $\mathbb{C}^2$ are confined to a union Σ of characteristic complex lines $\{z = z_0\}$, $\{w = w_0\}$ that are tangent to the variety Γ (of course, at characteristic points of Γ) on which the (CP) is set.

EXAMPLE 13.1. Let $\mathcal{L} := \Delta$.

(i) Take $\gamma = \mathbb{R}$, then $S(z) = z$. Thus, all solutions of (13.3) are entire harmonic functions (already noted in connection with the Bony-Schapira theorem — cf. Corollary 8.3).

(ii) Take $\gamma = \{z : |z| = R\}$, then $S(z) = \frac{R^2}{z}$. Thus, all solutions of (13.3) extend harmonically to $\mathbb{R}^2 \setminus \{0\}$.

However, solutions can become multi-valued when continued in the complex space, e.g., for $R = 1$, the solution u_Γ defined by (13.4) is given by $u_\Gamma = 2\log|z| + 1$ and is multi-valued in $\mathbb{C}^2$.

(iii) Consider any quadratic curve, e.g., an ellipse,

$$\left\{(x, y) : \frac{x^2}{a^2} + \frac{y^2}{b^2} = 1, a > b, a^2 - b^2 = 1\right\}.$$

Then $S(z) = \left(a^2 + b^2\right) z - 2ab\sqrt{z^2 - 1}$ and all solutions of (13.3) extend harmonically to $\mathbb{R}^2 \setminus \{\pm 1\}$.

1. Analytic continuation of potentials and quadrature domains

Recall the Herglotz problem stated in Sec. 1 of Ch. 1 but now posed in $\mathbb{R}^2$. Consider a domain Ω bounded by a simple analytic curve γ and define the potential of Ω by

$$u_\Omega(z) = \frac{1}{2\pi}\int_\Omega \log|\xi - z|\, dA(\xi).$$

(dA is the Lebesgue measure.) Then u_Ω solves (in the distributional sense) the problem $\Delta u_\Omega = \chi_\Omega$ and as is easily seen is C^1-smooth in $\mathbb{R}^2$. The problem is then to find a harmonic extension of u_Ω across γ into Ω. Let u_γ be the solution of the (CP)

$$\begin{cases} \Delta u_\gamma = 0, \\ \partial^\alpha \left(u_\gamma - \frac{1}{4}|z|^2\right) = 0,\ |\alpha| \le 1 \text{ on } \gamma. \end{cases} \tag{13.6}$$

Let us denote by $v_\gamma := \frac{1}{4}|z|^2 - u_\gamma$ the solution of

$$\begin{cases} \Delta v_\gamma = 1, \\ \partial^\alpha v_\gamma = 0,\ |\alpha| \le 1 \text{ on } \gamma. \end{cases} \tag{13.7}$$

PROPOSITION 13.1. *$u_\Omega(z) - v_\gamma(z)$ is harmonic in Ω and provides the desired extension of u_Ω inside Ω.*

PROOF. In Ω and near γ we have $\Delta(u_\Omega - v_\gamma) = 1 - 1 = 0$, and on γ, $\partial^\alpha(u_\Omega - v_\gamma) = \partial^\alpha u_\Omega$ for $|\alpha| \le 1$. Then we have the functions u_Ω and $u_\Omega - v_\Gamma$ harmonic in $\mathbb{R}^2 \setminus \Omega$ and in Ω, respectively, and having the same Cauchy data on γ. Hence, by Kovalevskaya's theorem (Theorem 6.1), $u_\Omega - v_\gamma$ provides the desired extension.

We consider now the above problem in greater generality. Define

$$u(z) = \frac{1}{2\pi}\int_\Omega p(\xi)\log|\xi - z|\,dA(\xi),$$

where $p(\xi)$ is, say, a polynomial (or an entire function) of two real variables.

Let P be another polynomial such that $\Delta P = p$ and $u_{\gamma,P}$ be a solution of

$$\begin{cases} \Delta u_{\gamma,P} = 0, \\ \partial^\alpha(u_{\gamma,P} - P) = 0,\ |\alpha| \le 1 \text{ on } \gamma. \end{cases} \tag{13.8}$$

Define $v_{\gamma,P} = P - u_{\gamma,P}$. It solves the problem

$$\begin{cases} \Delta v_{\gamma,P} = p, \\ \partial^\alpha v_{\gamma,P} = 0,\ |\alpha| \le 1 \text{ on } \gamma. \end{cases} \tag{13.9}$$

□

Then arguing as in the proof of Proposition 13.1 we obtain the following.

PROPOSITION 13.2. *The function $u_{\Omega,p}(z) - v_{\gamma,P}(z)$ is harmonic in Ω and provides the desired extension of $u_{\Omega,p}$ inside Ω.*

As a corollary of this result and Corollary 13.1 we obtain Herglotz's theorem mentioned in the introduction (Sec. 1 of Ch. 1).

COROLLARY 13.2 (Herglotz's theorem). *All potentials $u_{\Omega,p}$ with polynomial, or even entire densities extend harmonically into Ω along any path free of singularities of the Schwarz function of the boundary γ of Ω or, equivalently, free of singularities of u_γ the solution of the Cauchy problem*

$$\begin{cases} \Delta u_\gamma = 0; \\ \partial^\alpha\left(u_\gamma - \frac{1}{4}|z|^2\right) = 0,\ |\alpha| \le 1 \text{ on } \gamma. \end{cases} \tag{13.10}$$

The function u_γ turns out to be responsible for a great many potential-theoretic properties of the domain Ω.

In particular, let us note in passing the following.

REMARK 13.2. Suppose the Schwarz function S of γ extends as a distribution inside Ω. Since it is analytic near γ, in the sense of distributions we have $\frac{\partial S}{\partial \bar z} = T$, where T is a distribution compactly supported in Ω. Then the following quadrature identity holds:

$$\int_\Omega f\,dA = \int_\Omega f\,dT = \langle f, T\rangle$$

for all analytic functions f in $\overline{\Omega}$.

Indeed, by Green's formula

$$\int_\Omega f\,dA = \frac{1}{2i}\int_\gamma f\,\bar z\,dz = \frac{1}{2i}\int_\gamma f\,S(z)\,dz = \int_\Omega f\,\frac{\partial S}{\partial \bar z}\,dA = \langle f, T\rangle.$$

Similarly, if $\Delta u_\gamma = T$, we have that the quadrature identity

$$\int_\Omega v\, dA = \langle v, T\rangle$$

holds for all v harmonic in $\overline{\Omega}$.

Ω is then called a *quadrature domain* with respect to the distribution T. (In the context of Chapters 21 and 22, such Ω will be referred to as a *quadrature domain in the wide sense.*)

For example, if Ω is the interior of an ellipse

$$\gamma := \left\{(x, y) \in \mathbb{R}^2 : \frac{x^2}{a^2} + \frac{y^2}{b^2} = 1, a > b, a^2 - b^2 = 1\right\},$$

then (cf. Corollary 15.2)

$$\int_\Omega v\, dA = \text{const} \cdot \int_{-1}^{1} v(t)\sqrt{1-t^2}\, dt$$

for all harmonic polynomials v, where we leave it as an exercise for the reader to find the value of the constant in the above quadrature formula.

2. The Schwarz potential conjecture

In order to investigate a possibility of extending Corollaries 13.1 and 13.2 to higher dimensions, let us give the following definition.

DEFINITION 13.1. We shall call the Schwarz Potential (SP) u_Γ of a non-singular, real analytic hypersurface $\Gamma \subset \mathbb{R}^n$, $n \geq 2$ the solution of the (CP)

$$\begin{cases} \Delta u_\Gamma = 0; \\ \partial^\alpha\left(u_\Gamma - \frac{1}{2}\|x\|^2\right) = 0,\ |\alpha| \leq 1 \text{ on } \Gamma. \end{cases} \tag{13.11}$$

The following conjecture, if true, would be an analogue of Corollary 13.1.

THE (SP)-CONJECTURE. All solutions of the Cauchy problems

$$\begin{cases} \Delta u = 0; \\ \partial^\alpha(u - f) = 0,\ |\alpha| \leq 1 \text{ on } \Gamma \end{cases}$$

with entire data f, extend harmonically along all paths free of singularities of u_Γ.

For the sake of convenience, let us adopt the following notation.

NOTATION. We shall write $f \equiv g$ on Γ for $\partial^\alpha(g - f) = 0$, $|\alpha| \leq 1$ on Γ.

To see why the (SP)-Conjecture may be at least plausible, let us note the following proposition.

PROPOSITION 13.3. *Let $\widehat{\Gamma}$ be, as usual, the complexification of Γ, i.e.,*

$$\widehat{\Gamma} = \{z \in \mathbb{C}^n : \phi(z) = 0\},$$

where ϕ is holomorphic and non-singular in a neighborhood of Γ and such that $\widehat{\Gamma} \cap \mathbb{R}^n = \Gamma$. Let $z^0 \in \widehat{\Gamma}$ be a characteristic point (with respect to Δ). Then u_Γ must be singular at z^0.

PROOF. Suppose u_Γ is holomorphic near z^0. Then, so is $v_\Gamma := \frac{1}{2}\sum_{j=1}^{n} z_j^2 - u_\Gamma$.

Now, $v_\Gamma \equiv 0$ on $\widehat{\Gamma}$ and $\Delta v_\Gamma = n \neq 0$. Since $v_\Gamma \equiv 0$ on $\widehat{\Gamma}$, the Weierstrass Preparation Theorem [**66**, Ch. 0, Sec. 1] implies that $v_\Gamma = \phi^2 g$ for some g holomorphic near z^0.

But on the other hand since $\phi\left(z^0\right) = 0$ we have

$$n = \Delta\left(\phi^2 g\right)\big|_{z^0} = 2\sum_{j=1}^{n}\left(\frac{\partial\phi}{\partial z_j}\right)^2 g|_{z^0}.$$

This gives a desired contradiction since z^0 is a characteristic point and hence $\sum_{j=1}^{n}\left(\frac{\partial\phi}{\partial z_j}\right)^2\bigg|_{z^0} = 0$. □

According to an intuitive feeling that singularities of solutions to a Cauchy problem appear only as a result of propagation through $\mathbb{C}^n$ of singularities originated on the initial surface, where by the above proposition the (SP) picks up all of them, we "conclude" that the (SP) should in principle exhibit all possible singularities for solutions of all Cauchy problems (for the Laplace equation).

REMARK 13.3. If this conjecture was indeed true it would, in particular, have answered Herglotz' question in all dimensions. Namely, the same word-for-word argument as that used in two dimensions would show that all potentials $u_{\Omega,p}$ with a polynomial (or even an entire) density p extend inside Ω to any region free of singularities of $u_{\partial\Omega}$ (the (SP) of the boundary $\partial\Omega$ of Ω).

It is easy to check that Proposition 13.3 extends to any linear operator $\mathcal{L} = \sum_{|\alpha|\le m} a_\alpha \partial^\alpha$ with holomorphic coefficients. Namely, suppose $\Gamma := \{\varphi(z) = 0\}$ with φ holomorphic near Γ, and $\nabla\varphi \neq (0,..,0)$ on Γ, and consider the solution u_Γ of the (CP)

$$\left\{\begin{array}{l}\mathcal{L}u_\Gamma = 1,\\ \partial^\alpha u_\Gamma = 0, |\alpha| \le m-1 \text{ on } \Gamma.\end{array}\right. \tag{13.12}$$

The solution u_Γ must be singular at *all* characteristic (with respect to $\mathcal{L}$) points of Γ, i.e., at all point z^0 satisfying

$$\varphi(z^0) = 0, \qquad \sum_{|\alpha|=m} a_\alpha(z^0)(\nabla_z\varphi(z^0))^\alpha = 0.$$

This follows as before since if u_Γ were holomorphic near z^0, then $u_\Gamma = \varphi^m f$ and accordingly we would have

$$1 = \mathcal{L}u_\Gamma = m\left(\sum_{|\alpha|=m} a_\alpha(\nabla_z\varphi)^\alpha\right) f + \varphi(\text{holomorphic function}).$$

If f is holomorphic at a characteristic point z^0, the right hand side vanishes at z^0 while the left hand side does not. Hence, one might ask whether in general, for any operator $\mathcal{L}$ with entire coefficients all solutions of all (CP) with entire data extend holomorphically to any domain free of singularities of the solution u_Γ of (13.12). The following example [**102**] shows that without additional assumptions on $\mathcal{L}$, this is false. Let

$$\mathcal{L} := \frac{\partial}{\partial z_2} - z_1^2\frac{\partial}{\partial z_1}, \qquad \Gamma := \{z_2 = 0\}.$$

Then $u_\Gamma = z_2$ is entire (note that $\mathcal{L}$ is of the first order, so u_Γ only needs to vanish to zeroth order along Γ). Yet, *any* solution of the (CP)

$$\begin{cases} \mathcal{L}u = 0, \\ u(z_1, 0) = f(z_1), \end{cases} \tag{13.13}$$

with f entire has the form $u(z_1, z_2) = f\left(\frac{z_1}{1 - z_1 z_2}\right)$ and, in general, develops singularities on the same characteristic variety $\{z_1 z_2 = 1\}$ tangent to Γ at infinity.

NOTE 13.1. (i) It is crucial in the above example that the coefficient of $\partial/\partial z_1$ is z_1^2 and not z_1. Indeed for $\mathcal{L} = \partial/\partial z_2 - z_1 \partial/\partial z_1$, $\Gamma = \{z_2 = 0\}$, $u_\Gamma = z_2$ and all solutions of (13.13) with entire data f have the form $f(z_1 e^{z_2})$, i.e., such solutions are entire functions (see Ch. 20 for a detailed treatment of equations of that type, Persson equations; cf. [**94**]).

(ii) Corollary 13.2 (as noted in [**102**]) shows that for arbitrary second order operators of two variables with constant coefficients in the principal part, the extended Schwarz potential conjecture nevertheless holds.

The following simple example is the first in a row of similar examples continued in the next sections for which the (SP)-Conjecture does hold.

EXAMPLE 13.2. Suppose $\Gamma = \left\{ x \in \mathbb{R}^n : \sum_{j=1}^{n} a_j x_j - c = 0 \right\}$ is a hyperplane. First, let us find u_Γ. It is easier to start out with $v_\Gamma = \frac{1}{2}\|x\|^2 - u_\Gamma$ that solves

$$\begin{cases} \Delta u_\Gamma = n, \\ v_\Gamma \equiv 0 \text{ on } \Gamma. \end{cases}$$

One easily guesses that

$$v_\Gamma = \frac{n}{\|a\|^2} \left(\sum_{j=1}^{n} a_j x_j - c \right)^2,$$

where $a = (a_1, \ldots, a_n)$. Hence,

$$u_\Gamma = \frac{1}{2}\|x\|^2 - \frac{n}{2\|a\|^2} \left(\sum_{j=1}^{n} a_j x_j - c \right)^2.$$

The (SP)-Conjecture then would imply that solutions to all Cauchy problems with entire data on Γ are in fact entire harmonic functions; this is indeed the case in view of the Bony-Schapira theorem (cf. Corollary 8.3) since every characteristic (with respect to the Laplacian) complex hyperplane intersects $\widehat{\Gamma}$, the complexification of Γ. Another more direct proof of this can be obtained by applying the estimates in Chapter 2; cf. Exercise 3.3.

REMARK 13.4. Of course, the previous example can be extended to the case of a complex hyperplane $\widehat{\Gamma} = \left\{ z \in \mathbb{C}^n : \sum_{j=1}^{n} a_j z_j - c = 0 \right\}$ as long as $\widehat{\Gamma}$ is not characteristic, i.e., $\sum_{j=1}^{n} a_j^2 \neq 0$. In fact, Proposition 13.3 shows that for characteristic $\widehat{\Gamma}$, $u_{\widehat{\Gamma}}$ does not even exist.

REMARK 13.5. A result in [**102**] shows that the SP-conjecture for $\mathcal{L} = \sum_{j=1}^{3} \frac{\partial^2}{\partial z_j^2}$ holds for all polynomial data for curious "cylinders" in $\mathbb{C}^3$ given by

$$\Gamma := \{z \in \mathbb{C}^3 : z_3 = \varphi(z_1 + iz_2)\},$$

where φ is a holomorphic function of one variable. One easily verifies that $v_\Gamma = \frac{1}{2}[\varphi(z_1 + iz_2) - z_3]^2$. It is shown by an inductive argument in [**102**] that for any polynomial p, the solution of the (CP)

$$\begin{cases} \mathcal{L}u = 0, \\ \partial^\alpha(u - p) = 0 \text{ on } \Gamma, |\alpha| \leq 1, \end{cases}$$

extends to any region free of singularities of u_Γ. In particular, if φ is an entire function, all solutions are entire. It is not known whether the latter result holds for all entire data though.

Notes

The presentation in this chapter followed closely the lecture notes [**103**] of the first author. The important yet simple Theorem 13.1 is contained in various forms in [**57**], [**81**], [**131**], [**196**]. Here we followed the presentation in [**102**], where the reader can find more references. Most frequently, it is stated in the form of Corollary 13.1 (first statement) — cf., e.g., [**31**]. Propositions 13.1 and 13.2 and Corollary 13.2 are from Herglotz' often overlooked memoir [**83**]. Yet some versions of it, in all dimensions, were noted earlier by E. Schmidt and later by R. Wavre. We refer the reader to a detailed historic account and references in Shapiro's book [**176**].

The remark following Corollary 13.2 is a subject on its own. Someone interested in pursuing this direction further is referred to works by D. Aharonov, B. Gustafsson, M. Sakai, H.S. Shapiro, and references therein. Here we simply refer the reader to [**31**], [**110**], [**165**], [**176**], [**70**] where a great many references are given. Definition 13.1 and the (SP)-Conjecture appeared in [**110**] as did Proposition 13.3 although, to some degree, the (SP)-Conjecture was anticipated in [**174**]. Also, cf. [**102**] and [**176**]. The first few nontrivial cases (spheres, cylinders, cones and ellipsoids) for which the (SP)-Conjecture holds although restricted to polynomial data were investigated in [**110**] by ad-hoc, "hands on" methods. Some of those results are discussed in the following sections. There is by now considerable progress achieved in verifying the Conjecture mostly due to the efforts of Sternin and Shatalov and their school (cf., e.g., their book [**181**]) and, independently, by G. Johnsson [**94**]. The latter, in particular, verified the conjecture for all quadratic surfaces.

A direct treatment of Example 13.2 can be found, for instance, in Hörmander's book [**86**, Lemma 9.1.4].

The topic of quadrature domains is discussed further in Chapters 21 and 22.

Remark 13.1 stems from a deep theory of J. Leray [**126**] discussed in Ch. 19.

CHAPTER 14

The Schwarz Potential Conjecture for Spheres

Note to the reader: This chapter is heavier on various indices, so to avoid a notational hazard we have used "N" instead of "n" for the dimension of the space in this chapter and in some of the following chapters as well.

Let $\Gamma = \{x \in \mathbb{R}^N : |x| = R\}$. First let us calculate the SP u_Γ by solving

$$\begin{cases} \Delta u_\Gamma = 0, \\ u_\Gamma \equiv \frac{1}{2}|x|^2 \text{ on } \Gamma. \end{cases}$$

The rotational symmetry of Δ, Γ and the data together with uniqueness of solutions imply that one should seek a solution in the form

$$u_\Gamma = \begin{cases} c_1 \log|x| + c_2, & \text{when } N = 2, \\ c_1|x|^{2-N} + c_2, & \text{when } N \geq 3. \end{cases}$$

Calculating the constants we find

$$u_\Gamma = \begin{cases} R^2\left(\log|x| + \frac{1}{2} - \log R\right), & \text{when } N = 2, \\ -\dfrac{R^N}{N-2}\dfrac{1}{|x|^{N-2}} + \dfrac{N}{2(N-2)}R^2, & \text{when } N \geq 3. \end{cases}$$

So the SP conjecture would imply the following

THEOREM 14.1. *All solutions of the CP*

$$\begin{cases} \Delta u = 0, \\ u \equiv f \text{ on } \Gamma \end{cases}$$

with entire data f extend harmonically to all of $\mathbb{R}^N \setminus \{0\}$.

Recall that "$u \equiv f$ on Γ" means that u and f as well as their gradients coincide on Γ.

The case $N = 2$ is covered in the previous chapter, so from now on, we shall assume $N \geq 3$ and $R = 1$.

Our strategy is rather straightforward.

Let $P_m :=$ {polynomials in N variables of degree $\leq m$} and let $H_k \subset P_k$ be a subspace of homogeneous polynomials of degree k.

Let $f(x) = \sum_{m=0}^{\infty} f_m(x)$, where $f_m \in H_m$.

We are going to solve the (CP)

$$\begin{cases} \Delta U_m = 0, \\ U_m \equiv f_m \text{ on } \Gamma \end{cases}$$

for all $m \geq 0$ and show that the series $\sum_{m=0}^{\infty} U_m(x)$ converges uniformly on compact subsets of $\mathbb{R}^N \setminus \{0\}$.

LEMMA 14.1. *Let $f_m \in P_m$. Then, $f_m \equiv u_m + \left(|x|^2 - 1\right) v_m$ on Γ, where $u_m \in P_m$, $v_m \in P_{m-2}$ (or $= 0$ if $m = 0, 1$) are harmonic polynomials.*

PROOF. Consider operator $T : P_m \to P_m$ defined by

$$T(p) = \Delta\left[\left(|x|^2 - 1\right) p\right].$$

CLAIM. T is surjective.

Since $\dim P_m < +\infty$ it is sufficient to check that T is injective. Let $p \in Ker\, T$, then $\left(|x|^2 - 1\right) p$ is harmonic. But it vanishes on Γ, hence the maximum principle implies that it vanishes identically. Therefore $p \equiv 0$.

For $m = 0, 1$, the lemma is obvious, so let $m \geq 2$. Find $p \in P_{m-2}$ such that $\Delta\left[\left(|x|^2 - 1\right) p\right] = \Delta f_m$. Then $u_m := f_m - \left(|x|^2 - 1\right) p \in P_m$ is harmonic. Now by the same token $p = v_m + \left(|x|^2 - 1\right) q$, where $v_m \in P_{m-2}$ is harmonic and $q \in P_{m-4}$. So,

$$f_m = u_m + \left(|x|^2 - 1\right) v_m + \left(|x|^2 - 1\right)^2 q$$

and the lemma follows. □

Let

$$\text{(14.1)} \qquad \begin{aligned} u_m &= u_{m,0} + u_{m,1} + \cdots + u_{m,m}, \\ v_m &= v_{m,0} + v_{m,1} + \cdots + v_{m,m-2} \end{aligned}$$

denote the decomposition of u_m and v_m into homogeneous polynomials; thus $u_{m,j}$, $v_{m,j}$ are harmonic and in H_j.

LEMMA 14.2. *The solution U_m of the (CP)*

$$\begin{cases} \Delta U_m = 0, \\ U_m \equiv f_m \ \text{on } \Gamma, \quad f_m \in H_m \end{cases}$$

is given by

$$\text{(14.2)} \qquad U_m = \sum_{k=0}^{m} u_{m,k} + \sum_{k=0}^{m-2} \frac{2}{2 - N - 2k}\left(\frac{v_{m,k}}{|x|^{N-2+2k}} - v_{m,k}\right),$$

where $u_{m,k}$ and $v_{m,k}$ are the same as in (14.1) (and Lemma 14.1). (In the trivial cases of $m = 0, 1$, $v_{m,k} = 0$.)

PROOF. Recall Euler's formula: if $g \in H_k$, then

$$\text{(14.3)} \qquad \sum_{1}^{N} x_j \frac{\partial g}{\partial x_j} = kg.$$

Let $h \in H_k$ be harmonic. We have

$$\begin{aligned} \frac{\partial^2}{\partial x_j^2}\left(\frac{h}{\left(|x|^{N-2-2k}\right)}\right) &= h_{x_j x_j} |x|^{N-2-2k} + 2(2 - N - 2k) x_j |x|^{-N-2k} h_{x_j} \\ &\quad - (2 - N - 2k)(N + 2k)\, h x_j^2 |x|^{-N-2-2k} \\ &\quad + (2 - N - 2k)\, |x|^{-N-2k}\, h. \end{aligned}$$

Summing over j and using (14.3) we conclude that $\dfrac{h}{|x|^{N-2+2k}}$ is harmonic. Now $\sum_{k=0}^{m} u_{m,k} = u_m$ while

$$\frac{2}{2-N-2k}\left(\frac{v_{m,k}}{|x|^{N-2-2k}} - v_{m,k}\right) \equiv \left(|x|^2 - 1\right) v_{m,k} \text{ on } \Gamma. \tag{14.4}$$

(14.4) and Lemma 14.1 complete the proof of Lemma 14.2. $\square$

It's worthwhile to pause here to observe the following corollary.

COROLLARY 14.1. *All solutions of CP with polynomial data on the sphere Γ extend harmonically to $\mathbb{R}^N \setminus \{0\}$ thus confirming the SP conjecture for polynomial data.*

REMARK 14.1. With a little bit more work (we leave it as an exercise), one can show that any solution U_m of the (CP) with data $f_m \in P_m$ can be written in the form

$$U_m = u_m + R(D)\,(u_\Gamma)\,.$$

where $u_m \in P_m$ is harmonic and R is a polynomial.

LEMMA 14.3. *Let*

$$F_m := max\{|f_m(x)| : x \in \Gamma\}\,,$$

and

$$G_m := max\{\|\nabla f_m(x)\| : x \in \Gamma\}\,.$$

Then,

$$\lim_{m\to\infty} (F_m)^{\frac{1}{m}} = \lim_{m\to\infty} (G_m)^{\frac{1}{m}} = 0.$$

PROOF. We shall conduct the argument for G_m, the proof for F_m is similar.

For any $x \in \Gamma$ fixed and any $t \in \mathbb{C}$, the function $t \to f(tx)$ is an entire analytic function.

We have

$$f(tx) = \sum_0^\infty t^m f_m(x).$$

For each j, $1 \le j \le N$, we have

$$\frac{\partial f(tx)}{\partial x_j} = \sum_0^\infty t^m \frac{\partial f_m(x)}{\partial x_j},$$

i.e., $\dfrac{\partial f_m(x)}{\partial x_j}$ are the Taylor coefficients of the entire function $t \to \dfrac{\partial f(tx)}{\partial x_j}$.

The Cauchy-Hadamard estimate then implies

$$\left|\frac{\partial f_m(x)}{\partial x_j}\right| \le \frac{\max\left\{\left|\frac{\partial f(tx)}{\partial x_j}\right| : |t| \le T\right\}}{T^m}$$

for all $T > 0$. Therefore,

$$\max\left\{\left|\frac{\partial f_m(x)}{\partial x_j}\right| : x \in \Gamma\right\} \le \frac{\max\left\{\left|\frac{\partial f(z)}{\partial x_j}\right| : z \in \mathbb{C}^N, \|z\| \le T\right\}}{T^m}$$

and

$$\max\{\|\nabla f_m(x)\| : x \in \Gamma\} \le \frac{\left(\sum_{j=1}^{N}\left\{\max\left\{\left|\frac{\partial f(z)}{\partial x_j}\right| : \|z\| \le T\right\}\right\}^2\right)^{\frac{1}{2}}}{T^m}.$$

Taking the mth-root and letting $m \to \infty$ gives

$$\overline{\lim_{m\to\infty}} G_m^{\frac{1}{m}} \le \frac{1}{T}.$$

Since T was arbitrary, the lemma follows. □

From now on A_N, B_N, C_N, etc., denote constants that only depend on the dimension of the space.

LEMMA 14.4. *Let $h \in P_m$ be harmonic and $h = h_0 + h_1 + \cdots + h_k + \cdots + h_m$ be its decomposition into homogeneous harmonic polynomials. Then,*

$$\max_{x\in\Gamma} |h_k(x)| \le C_N k^{\frac{N}{2}} \max_{x\in\Gamma} |h(x)|, \quad 1 \le k \le m.$$

Also,

$$|h_0(x)| = |h_0(0)| \le \max_{x\in\Gamma} |h(x)|.$$

PROOF. Without loss of generality suppose $\max_{x\in\Gamma} |h(x)| = 1$. The case $k = 0$ is obvious, so consider $k \ge 1$.

Recall that $h_i \perp h_j$ if $i \ne j$ in $L^2(\Gamma, d\sigma)$, where $d\sigma$ is the surface measure on Γ. (A quick outline of a proof is provided in Exercise 14.1 at the end of this chapter.) Hence

$$\int_\Gamma |h_k|^2\, d\sigma \le \int_\Gamma |h|^2 d\sigma \le |\Gamma|,$$

where $|\Gamma|$ is the $(N-1)$-dimensional measure of Γ.

Therefore, it follows easily that (exercise)

$$\int_B |h_k|^2\, d\sigma \le A_N, \tag{14.5}$$

where $B := \{x : |x| \le 1\}$ is the unit ball.

Fix $y \in \Gamma$. Then, for $0 < r < 1$, the mean value theorem applied to the ball B' centered at ry with radius $(1-r)$ and the subharmonic function $|h_k|^2$ gives the estimate

$$|h_k(ry)|^2 \le \frac{1}{|B'|}\int_{B'} |h_k(x)|^2\, dx \le \frac{A_N}{A'_N(1-r)^N} = \frac{A''_N}{(1-r)^N}.$$

From homogeneity of h_k we then derive

$$|h_k(y)| \le \left[\frac{A''_N}{r^{2k}(1-r)^N}\right]^{\frac{1}{2}}$$

for all $0 < r < 1$. Taking $r = 1 - \frac{1}{2k}$ we obtain the lemma. □

LEMMA 14.5. *Let $f_m \equiv u_m + \left(|x|^2 - 1\right) v_m$ on Γ be as in Lemma 14.1. Then*

$$V_m := \max_{x\in\Gamma} \{|v_m(x)|\} \le C_N \left(G_m + m^{2N} F_m\right),$$

where $F_m = \max_{x\in\Gamma}\{|f_m(x)|\}$, $G_m = \max_{x\in\Gamma}\{\|\nabla f_m(x)\|\}$ are the same as in Lemma 14.3. In particular, $\lim_{m\to\infty} V_m^{\frac{1}{m}} = 0$.

PROOF. On Γ by the hypothesis we have for $1 \leq j \leq N$

$$\frac{\partial f_m}{\partial x_j} = \frac{\partial u_m}{\partial x_j} + 2x_j v_m(x).$$

Therefore,

$$4\sum_1^N x_j^2 \left|v_m\right|^2 = 4\left|v_m\right|^2 \leq 2\left(\left\|\nabla f_m\right\|^2 + \left\|\nabla u_m\right\|^2\right)$$

on Γ, i.e., for $x \in \Gamma$,

$$|v_m(x)| \leq C\left(G_m + \|\nabla u_m(x)\|\right). \tag{14.6}$$

LEMMA 14.6. *Let $h \in H_k$ be a homogeneous polynomial of degree k. Then*

$$\max_{x\in\Gamma} \|\nabla h(x)\| \leq k\sqrt{2}\max_{x\in\Gamma}|h(x)|.$$

Assume Lemma 14.6 for the moment.

From Lemmas 14.4, 14.6 and the fact that $u_m = f_m$ on Γ, we obtain for $x \in \Gamma$

$$\left|\frac{\partial u_m}{\partial x_j}\right| \leq \sum_{k=0}^m C_N k^{\frac{N}{2}+1} F_m$$

and, finally,

$$\|\nabla u_m(x)\| \leq C_N m^{\frac{N+4}{2}} F_m \leq C_N m^{2N} F_m. \tag{14.7}$$

Now, (14.6) and (14.7) imply Lemma 14.5. □

PROOF OF LEMMA 14.6. Fix $x \in \Gamma$. Then, in view of (14.3), the normal derivative of h at x equals

$$\frac{\partial h}{\partial n}(x) = \sum_1^N x_j \frac{\partial h}{\partial x_j} = kh(x). \tag{14.8}$$

Fix vector $\vec{y}$, $\|\vec{y}\| = 1$ tangent to Γ at x. Consider the 2-dimensional plane Π spanned by vectors $\vec{x}$, $\vec{y}$. Π intersects Γ over a unit circle T. h restricted to the plane Π is a homogeneous polynomial H_k of degree k. In particular, by Lemma 14.1, the restriction of h on T, $H_k\big|_T$, is a harmonic polynomial of two variables of degree $\leq k$. Thus, if we use polar coordinates (r, θ) in the plane Π, we have on T

$$H_k\big|_T = \sum_0^k (a_j \cos j\theta + b_j \sin j\theta) = H_k(\theta),$$

i.e., H_k becomes a trigonometric polynomial of order k.

Then, invoking Bernstein's classical inequality [**187**, Sec. 1.22] , we obtain

$$\begin{aligned} |D_{\vec{y}}h(x)| &= \left|\left.\frac{dH_k}{d\theta}\right|_x\right| \leq k\max_{z\in T}|H_k(z)| \\ &= k\max_{z\in T}|h(z)| \leq k\max_{z\in\Gamma}|h(z)| \end{aligned} \tag{14.9}$$

for an arbitrary vector $\vec{y}$ tangent to Γ at x. From (14.8) and (14.9) the lemma follows. □

REMARK 14.2. Although it does not matter for the sequel one can easily drop the $\sqrt{2}$ in the above Lemma 14.6 in order to obtain the sharp constant. In other words, the maxima of a normal and tangential derivative cannot occur at the same point. The following argument was kindly provided to us by V. Totik (written communication). We shall sketch it here while leaving the details to the interested reader as an exercise.

Assume that our polynomial h has norm 1. Take any point Q on the sphere and any direction $\vec{D}$ at that point. Then there is a tangential direction $\vec{E}$ such that $\vec{D} = a\vec{E} + b\vec{R}$, where $\vec{R}$ is the radial direction (from the origin to Q) and

$$a^2 + b^2 = 1. \tag{14.10}$$

There is a plane containing the origin, Q, and the direction $\vec{E}$, and the intersection of that plane with the sphere is a circle T. On that circle (after introducing $\cos t$ and $\sin t$) the polynomial h becomes a trigonometric polynomial H_k of degree k, the derivative A of h in the direction $\vec{E}$ is just plus or minus $H_k'(t_0)$ (where the e^{it_0} point on the circle T corresponds to the point Q in question), while the radial derivative B of h at Q is $kh(Q) = kH_k(t_0)$. By the Szegő inequality (cf. [**190**])

$$A^2 + B^2 \leq k^2. \tag{14.11}$$

Now the derivative of h in the direction $\vec{D}$ is $aA + bB$ and is at most k in absolute value from (14.10) and (14.11) and the Cauchy-Schwarz inequality.

PROOF OF THEOREM 14.1. We want to show that $\sum_{m=0}^{\infty} |U_m(x)| \leq \infty$, for all x, $0 < |x| < \infty$, where U_m is defined by (14.2).

For that, it suffices to show that the series:

(I) $\sum\limits_{m=0}^{\infty} |u_m(x)|$,

(II) $\sum\limits_{m=0}^{\infty} \sum\limits_{k=0}^{m-2} \frac{1}{|2-N-2k|} \frac{|v_{m,k}(x)|}{|x|^{N-2+2k}}$, and

(III) $\sum\limits_{m=0}^{\infty} \sum\limits_{k=0}^{m-2} \frac{1}{|2-N-2k|} |v_{m,k}(x)|$ all converge.

(I) Lemma 14.3 implies that $F_m = \epsilon_m^m$, where $\lim \epsilon_m = 0$. Fix $x : \frac{1}{R} < |x| < R$, $R > 1$. From Lemma 14.4 it follows

$$|u_{m,k}| \leq C_N k^{\frac{N}{2}} \epsilon_m^m |x|^k \leq C_N' A_N^k \epsilon_m^m R^k, \tag{14.12}$$

where $A_N > 1$ is a constant. Thus,

$$\sum_{k=0}^{m} |u_{m,k}(x)| \leq C_N' \epsilon_m^m \sum_{k=0}^{m} (A_N R)^k \leq C_N'' \epsilon_m^m (A_N R)^{m+1} \tag{14.13}$$

and hence,

$$\sum_{m=0}^{\infty} \sum_{k=0}^{m} |u_{m,k}(x)| \leq C_N'' A_N R \sum_{m=0}^{\infty} (A_N \epsilon_m R)^m < A(R) < +\infty$$

because $\epsilon_m \to 0$ when $m \to \infty$.

It is worth pausing here to observe the following.

COROLLARY 14.2. *The solution* $u := \sum_{m=0}^{\infty} \sum_{k=0}^{m} u_{m,k}(x)$ *of the Dirichlet problem in the unit ball* B

$$\begin{cases} \Delta u = 0, \\ u = f \text{ on } \Gamma \end{cases}$$

with entire data f *is an entire harmonic function.*

We shall prove this result for general ellipsoids, not merely for spheres, in Chapter 17; many ideas from the current chapter are repeated there, but we treat it as a separate venue for the reader's convenience.

Note that the above argument immediately implies the convergence of (III) as well. Indeed, Lemma 14.5 implies that $V_m := \max |v_m(x)| = \delta_m^m$, $\delta_m \to 0$ while Lemma 14.3 provides the estimate

$$|v_{m,k}(x)| \leq C_N k^{\frac{N}{2}} \delta_m^m |x|^k \tag{14.14}$$

which is identical to (14.12).

Finally, to establish the convergence of (II), we fix $x : \frac{1}{R} < |x| < R$. Without loss of generality, we can assume $|x| < 1$, since for $|x| > 1$ convergence of (II) is implied by that of (III).

Then (14.14) yields (cf. (14.13))

$$\begin{aligned} \sum_{k=0}^{m-2} \frac{|v_{m,k}(x)|}{|x|^{N-2+2k}} &\leq C_N \delta_m^m \sum_{k=0}^{m-2} k^{\frac{N}{2}} R^{N-2+k} \\ &\leq C'_N \delta_m^m R^{N-2} (A_N R)^{m-1}. \end{aligned}$$

Therefore, as before

$$\sum_{m=0}^{\infty} \sum_{k=0}^{m-2} \frac{|v_{m,k}(x)|}{|x|^{N-2+2k}} < A(R) < +\infty$$

and hence series (II) converges as well. □

From the estimates we have given it follows at once that the series

$$u(x) := \sum_{m=0}^{\infty} U_m(x)$$

giving the solution of our Cauchy problem on the sphere converges absolutely everywhere in $\mathbb{C}^N \setminus \left\{z : \sum_1^N z_j^2 = 0\right\}$. Thus we obtain the following corollary establishing the (SP) conjecture for spheres in the $\mathbb{C}^N$-context.

COROLLARY 14.3. *The solution of the CP*

$$\begin{cases} \Delta u = 0, \\ u \equiv f \text{ on } \Gamma \end{cases}$$

with entire data f *extends (perhaps as a multi-valued function for odd* N*) to the whole complement in* $\mathbb{C}^N$ *of the isotropic cone* $\Gamma_0 := \left\{z : \sum_1^N z_j^2 = 0\right\}$.

EXERCISE 14.1 (Orthogonality of spherical harmonics). Suppose h_k and h_m are homogeneous harmonic polynomials of degree k and m, respectively, with $k < m$. Complete the following steps to show that h_k and h_m are orthogonal in the L^2 space of the unit sphere S^{N-1} (cf. [**180**]).

(a) Let $u(x)$ be a homogeneous polynomial (or even smooth homogeneous function) of degree j. Verify Euler's homogeneous function formula

$$\langle x, \nabla u(x) \rangle = j u(x).$$

Hint: Start by using the chain rule to differentiate $u(rx)$ with respect to a scalar r.

(b) Let ∂_n denote the normal derivative with respect to the outward pointing normal on the unit sphere. Then use part (a) to conclude that on S^{N-1} we have $\partial_n h_m(x) = m h_m(x)$ and $\partial_n h_k(x) = k h_k(x)$.

(c) Use part (b) to obtain

$$\int_{S^{N-1}} (m-k) h_m h_k = \int_{S^{N-1}} h_k \partial_n h_m - h_m \partial_n h_k.$$

Now apply Green's second identity to conclude that this integral vanishes, establishing the desired orthogonality.

Notes

Theorem 14.1 and Corollary 14.3 put in a context of the (SP) Conjecture are contained in G. Johnsson's Thesis [**94**]. In fact, Johnsson has even solved the problem for all second order operators that have the Laplacian as their principal part. Johnsson's work is rather deep and is based on globalizing families (as developed in Chapters 5 and 8) combined with the local uniformization of solutions of Cauchy problems pioneered by Leray [**126**] (see Chapters 19 and 20).

In another direction, generic results based on topological methods (e.g., R. Thom's theorem) have been developed by V. Shatalov, B. Sternin, and their school (see [**181**] and references therein).

The elementary proof of Theorem 14.1 presented here was found in [**104**]. Lemmas 14.1 and 14.2 and the following elementary Corollary 14.1 were obtained earlier in [**110**]. Lemmas 14.3 and 14.4 and Corollary 14.2 are contained in [**113**], where Corollary 14.2 is shown to hold for arbitrary ellipsoids. Finally, Lemmas 14.5 and 14.6, that tie the loose ends together, are from [**104**].

CHAPTER 15

Potential Theory on Ellipsoids: Part I - The Mean Value Property

The mean value property for harmonic functions can be rephrased as saying that *the average of any harmonic function over concentric balls is a constant.* As we formulate precisely below, there is a mean value property for ellipsoids which says *the average of any harmonic function over confocal ellipsoids is a constant* - cf. Fig. 1.

Consider a *heterogeneous* ellipsoid Γ (i.e., an ellipsoid with distinct semi-axes)

$$\Gamma = \left\{ x \in \mathbb{R}^N : \sum_{j=1}^{N} \frac{x_j^2}{a_j^2} - 1 = 0, a_1 > a_2 > \cdots \right\}$$

and let Ω be its interior.

DEFINITION 15.1. A family of ellipsoids $\{\Gamma_\lambda\}$,

$$\Gamma_\lambda = \left\{ x \in \mathbb{R}^N : \sum_{j=1}^{N} \frac{x_j^2}{a_j^2 + \lambda} - 1 = 0 \right\},$$

where $-a_N^2 < \lambda < +\infty$ is called a confocal family (for $N = 2$ these are ellipses with the same foci).

Observe that when $\lambda \to -a_N^2$,

$$\Gamma_\lambda \to \left\{ x \in \mathbb{R}^N : x_N = 0, \sum_{j=1}^{N-1} \frac{x_j^2}{a_j^2 - a_N^2} - 1 \leq 0 \right\} =: E.$$

E is called the *focal ellipsoid*.

The following classical theorem goes back to MacLaurin who considered prolate spheroids in $\mathbb{R}^3$ $(a_1 > a_2 = a_3)$. General ellipsoids were treated later by Laplace.

THEOREM 15.1. *Let u be, say, an entire harmonic function. Then*

$$\frac{1}{|\Omega_\lambda|} \int_{\Omega_\lambda} u(x)\, dx = \textit{constant} \tag{15.1}$$

for all $\lambda : \lambda > -a_N^2$.

From now on, for the sake of brevity, we shall only consider the case $N \geq 3$.

REMARK 15.1. MacLaurin's theorem can be deduced as a corollary of the following (1949) result of L. Asgeirsson [**11**] (cf. [**26**, Ch. VI, Sec. 16.1]).

"Suppose $u = u(x, y)$, where $x \in \mathbb{R}^{m_1}$, $y \in \mathbb{R}^{m_2}$ satisfy the ultrahyperbolic equation

$$\Delta_x u = \Delta_y u.$$

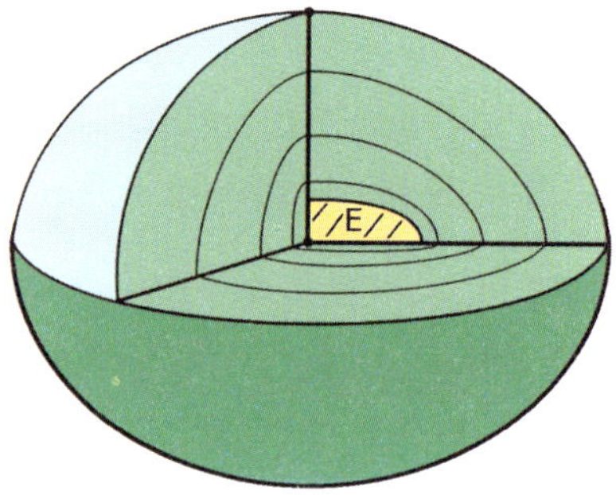

FIGURE 1. The mean value over confocal ellipsoids is constant.

Then if $\mu_i(x, y, r)$, $i = 1, 2$ denote respectively the mean values of u over m_i-dimensional balls of radius r centered at (x, y), we have $\mu_1(x, y, r) = \mu_2(x, y, r)$."

To see how the Asgeirsson theorem implies MacLaurin's, change variables to

$$x_i = \xi_i \cosh \alpha_i + \eta_i \sinh \alpha_i, \quad i = 1, \dots, N,$$

where α_i are some constants. One easily finds

$$\frac{\partial^2 u}{\partial \xi_i^2} = \frac{\partial^2 u}{\partial x_i^2} \cosh^2 \alpha_i,$$
$$\frac{\partial^2 u}{\partial \eta_i^2} = \frac{\partial^2 u}{\partial x_i^2} \sinh^2 \alpha_i.$$

Subtracting and summing over i we obtain $\Delta_\xi u = \Delta_\eta u$, and so we can apply Asgeirsson's theorem. MacLaurin's theorem follows at once by noting that spheres in the ξ and η variables respectively centered at the origin, i.e.,

$$\sum_{i=1}^N \frac{x_i^2}{\cosh^2 \alpha_i} \leq r^2$$

and

$$\sum_{i=1}^N \frac{x_i^2}{\sinh^2 \alpha_i} \leq r^2,$$

are nothing else but confocal ellipsoids in the x-variables.

1. Proof of MacLaurin's theorem using E. Fischer's inner product

Here we present a more self-contained proof of MacLaurin's theorem based on an elegant algebraic formalism; the following notions are due to E. Fischer [**52**]. Let H_k as before be the space of homogeneous polynomials of degree k. If $f \in H_k$, then

$$f(z) = \sum_{|\alpha|=k} f_\alpha z^\alpha.$$

Introduce an inner product on H_k (called the Fischer inner product), by letting

$$\langle z^\alpha, z^\beta \rangle = \begin{cases} 0, & \alpha \neq \beta, \\ \alpha!, & \alpha = \beta. \end{cases}$$

If $f = \sum_{|\alpha|=k} f_\alpha z^\alpha$, $g = \sum_{|\alpha|=k} g_\alpha z^\alpha$ then

$$\langle f, g\rangle = \sum_{|\alpha|=k} a! f_\alpha \overline{g_\alpha}.$$

Now consider the operator $\left(\frac{\partial}{\partial z}\right)^\alpha$ that maps H_k into $H_{k-|\alpha|}$, and $z^\alpha(f) = z^\alpha f$ which maps $H_{k-|\alpha|}$ back to H_k.

The main point of introducing such an inner product is the following:

Fact. The operators $\left(\frac{\partial}{\partial z}\right)^\alpha$ and z^α are adjoint with respect to $\langle , \rangle$.

We leave checking this property, as well as checking that $\langle , \rangle$ is indeed an inner product, as an exercise for the reader. The most immediate consequence is the following.

COROLLARY 15.1. *Let P be a polynomial with real coefficients. The operators $P(D)$ and multiplication by P are mutually adjoint with respect to $\langle , \rangle$.*

In the proof of MacLaurin's Theorem, we shall need the following lemma. Let $\mathcal{H}_m$ denote the subspace of homogeneous harmonic polynomials of degree m.

LEMMA 15.1. *The linear span of harmonic polynomials $\left(z \cdot \xi\right)^m$ for all $\xi \in \Gamma_0 = \left\{\xi \in \mathbb{C}^N : \sum_{j=1}^N \xi_j^2 = 0\right\}$ (the isotropic cone) equals $\mathcal{H}_m$.*

PROOF OF LEMMA 15.1. First, note that $\frac{1}{m!}\left(z \cdot \bar{\xi}\right)^m$ serves as a reproducing kernel for the space H_m, that is, for any polynomial $f \in H_m$, we have

$$\frac{1}{m!}\left\langle f, \left(z \cdot \bar{\xi}\right)^m\right\rangle = f(\xi). \tag{15.2}$$

Indeed, it suffices to check (15.2) for a monomial z^α with an arbitrary multi-index α; $|\alpha| = m$, and we have, using the multinomial formula and orthogonality (while using the multinomial coefficient notation $\binom{m}{\alpha} := \frac{m!}{\alpha!} = \frac{m!}{\alpha_1! \cdots \alpha_n!}$)

$$\begin{aligned}\frac{1}{m!}\left\langle z^\alpha, \left(z \cdot \bar{\xi}\right)^m\right\rangle &= \frac{1}{m!}\left\langle z^\alpha, \sum_{|\alpha|=m} \binom{m}{\alpha} z^\alpha \bar{\xi}^\alpha\right\rangle \\ &= \frac{1}{\alpha!}\left\langle z^\alpha, z^\alpha \bar{\xi}^\alpha\right\rangle = \xi^\alpha,\end{aligned}$$

and (15.2) follows.

Restricting to $\xi \in \Gamma_0$ as in the statement of the lemma, let us assume that $u \in \mathcal{H}_m$ satisfies

$$\left\langle u, \left(z \cdot \bar{\xi}\right)^m\right\rangle = 0, \quad \forall \xi \in \Gamma_0.$$

Then (15.2) implies $u(\xi) = 0$ for all $\xi \in \Gamma_0$. By Hilbert's Nullstellensatz [**121**, Ch. IX, Sec. 1], u is then divisible by the defining polynomial of Γ_0:

$$u(\xi) = \left(\sum_{j=1}^N \xi_j^2\right) q(\xi), \quad \text{for some } q \in H_{m-2}. \tag{15.3}$$

But then, since u is harmonic, we have (applying Corollary 15.1)

$$\begin{aligned} 0 &= \langle \Delta u, q \rangle \\ &= \left\langle u, \left(\sum_{j=1}^{N} \xi_j^2 \right) q \right\rangle \\ &= \langle u, u \rangle, \end{aligned}$$

where we have used (15.3) in the last equation. Thus, u must vanish identically. This proves the lemma. □

PROOF OF MACLAURIN'S THEOREM. It suffices to check (15.1) for harmonic homogeneous polynomials, and in view of Lemma 15.1, we just have to check it for polynomials

$$\left(z \cdot \bar{\xi} \right)^m, \quad \xi \in \Gamma_0.$$

Fix λ. Let $b_i = \left(a_i^2 + \lambda \right)^{1/2}$ be the semi-axes of Ω_λ. We have to show that

$$\frac{1}{|\Omega_\lambda|} \int_{\Omega_\lambda} \left(x \cdot \bar{\xi} \right)^m dx = \frac{1}{|\Omega|} \int_{\Omega} \left(y \cdot \bar{\xi} \right)^m dy, \quad \forall \xi \in \Gamma_0.$$

Changing variables in both integrals

$$x_k = a_k x_k'; \quad y_k = b_k y_k', \quad k = 1, ..., N,$$

we see that it suffices to show the following:

$$\int_B \left(\sum_{k=1}^{N} a_k x_k \overline{\xi_k} \right)^m dx = \int_B \left(\sum_{k=1}^{N} b_k x_k \overline{\xi_k} \right)^m dx,$$

where B is the unit ball in $\mathbb{R}^N$. Since for $\xi \in \Gamma_0$, we have

$$\sum_{k=1}^{N} (a_k \xi_k)^2 - \sum_{k=1}^{N} (b_k \xi_k)^2 = -\lambda^2 \sum_{k=1}^{N} \xi_k^2 = 0,$$

it suffices to check the following assertion.

Assertion. The polynomial

$$P(t) := \int_B \left(\sum_{k=1}^{N} t_k x_k \right)^m dx$$

depends only on $\sum_{k=1}^{N} t_k^2$, for $t \in \mathbb{C}^N$.

To prove the assertion first note that $P \in H_m$ and without loss of generality we can assume that $m > 0$.

Take $t \in \mathbb{R}^N$ and let U be an orthogonal transformation on $\mathbb{R}^N$. Then

$$P(Ut) = \int_B (Ut, x)^m dx = \int_B (t, U^* x)^m \, dx = \int_B (t, U^* x)^m \, d\left(U^* x \right) = P(t)$$

where $(,)$ is the usual Euclidean inner product, and U^* is the adjoint of U with respect to $(,)$.

Then $P(t) = C = const$ on the n-dimensional sphere S^{n-1}. Applying Hilbert's Nullstellensatz again we obtain

$$P(t) - C = \left(\sum_{j=1}^{N} t_j^2 - 1\right) Q(t),$$

where Q is some polynomial. But $P(0) = 0$, so $C = Q(0)$ and hence

$$P(t) = \left(\sum_{j=1}^{N} t_j^2\right)^r R(t),$$

where R is relatively prime to $\left(\sum_{j=1}^{N} t_j^2\right)$.

But R is constant on S^{N-1}, so repeating the above argument we conclude that $R(t)$ must be a constant. This proves our Assertion and hence the theorem. $\square$

2. The Newtonian potential of an ellipsoid

Let us return to the question of Herglotz in Ch. 1, Sec. 1 in the context of ellipsoids.

Let as above

$$\Omega = \left\{ x \in \mathbb{R}^N : \sum_{j=1}^{N} \frac{x_j^2}{a_j^2} \leq 1, a_1 > a_2 > \cdots > a_N > 0 \right\},$$

and consider the exterior potential

$$u_\Omega(x) := C_N \int_\Omega \frac{dy}{|x-y|^{N-2}}, \quad x \in \mathbb{R}^N \setminus \Omega$$

generated by uniform density on Ω. In the following, as in previous chapters, we use the notation $y' = (y_1, ..., y_{N-1})$.

COROLLARY 15.2. *For* $x \in \mathbb{R}^N \setminus \bar{\Omega}$

$$u_\Omega(x) = C_N \int_E \frac{d\mu(y)}{|x-y|^{N-2}},$$

where

$$d\mu(y) = 2\left(\prod_{j=1}^{N} a_j\right)\left(\prod_{j=1}^{N-1}(a_j^2 - a_N^2)\right)^{-1/2}\left(1 - \sum_{j=1}^{N-1} \frac{y_j^2}{a_j^2 - a_N^2}\right)^{1/2} dy' \, |_E$$

(dy' is Lebesgue measure on $\{y_N = 0\}$ and E denotes the focal ellipsoid as defined in the previous section).

PROOF. We have by MacLaurin's theorem

$$u_\Omega(x) = \lim_{\lambda \to -a_N^2} \frac{\prod_{j=1}^{N} a_j}{\prod_{j=1}^{N} \left(a_j^2 + \lambda\right)^{1/2}} \int_{\Omega_\lambda} v(y)\, dy,$$

where we set $v(y) : \frac{C_N}{|x-y|^{N-2}}$. Now

$$\frac{\prod_{j=1}^{N} a_j}{\prod_{j=1}^{N} \left(a_j^2 + \lambda\right)^{1/2}} \int_{\Omega_\lambda} v(y)\, dy$$

$$= \frac{\prod_{j=1}^{N} a_j}{\prod_{j=1}^{N} \left(a_j^2 + \lambda\right)^{1/2}} \left(\int_{\Omega_\lambda} \left(v(y) - v\left(y', 0\right)\right) dy + \int_{\Omega_\lambda} v\left(y', 0\right) dy \right).$$

We can apply Fubini's theorem to each of these integrals, and letting

$$A_\lambda\left(y'\right) = \left(1 - \sum_{j=1}^{N-1} \frac{y_j^2}{a_j^2 + \lambda}\right)^{1/2} \left(a_N^2 + \lambda\right)^{1/2},$$

we obtain

$$\begin{aligned} u_\Omega(x) = \lim_{\lambda \to -a_N^2} & \frac{\prod_{j=1}^{N} a_j}{\prod_{j=1}^{N} \left(a_j^2 + \lambda\right)^{1/2}} \left(\int_{\Omega_\lambda \cap \{y_N = 0\}} 2v\left(y', 0\right) A_\lambda\left(y'\right) dy \right. \\ & \left. + \int_{\Omega_\lambda \cap \{y_N = 0\}} \int_{-A_\lambda(y')}^{A_\lambda(y')} \left(v(y) - v\left(y', 0\right)\right) d_{y_N}\, dy' \right). \end{aligned} \tag{15.4}$$

Now, observing that

$$\left|v(y) - v\left(y', 0\right)\right| \leq \max_{\Omega} \|\nabla v\| \left(a_N^2 + \lambda\right)^{1/2},$$

we obtain from (15.4) that

$$u_\Omega(x) = \lim_{\lambda \to -a_N^2} \frac{\prod_{j=1}^{N} a_j}{\prod_{j=1}^{N} \left(a_j^2 + \lambda\right)^{1/2}} \int_{\Omega_\lambda \cap \{y_N = 0\}} 2v\left(y', 0\right) A_\lambda\left(y'\right) dy'$$

or, applying Lebesgue's dominated convergence theorem,

$$u_\Omega(x) = 2 \frac{\prod_{j=1}^{N} a_j}{\prod_{j=1}^{N-1} \left(a_j^2 - a_N^2\right)^{1/2}} \int_E \left(1 - \sum_{j=1}^{N-1} \frac{y_j^2}{a_j^2 - a_N^2}\right)^{1/2} v\left(y', 0\right) dy'$$

and the corollary follows. □

REMARK 15.2. For $N = 3$, with a little bit more care, the argument implies the statements made in the introduction (see Examples 1.2 and 1.3 in Ch. 1) concerning oblate and prolate spheroids. We leave it to the reader as a perhaps tedious but satisfying exercise.

Since the density of the distribution $d\mu$ is real analytic in the interior of E we note the following corollary:

COROLLARY 15.3. *The potential $u_\Omega(x)$ extends as a (multivalued) harmonic function into $\mathbb{R}^N \setminus \partial E$.*

Since inside Ω the Schwarz potential of the boundary Γ of Ω satisfies

$$u_\Gamma(x) = N\left(u_\Omega(x) - \hat{\mu}(x)\right) + \frac{1}{2}|x|^2$$

where $\hat{\mu}(x) := C_N \int_E \frac{d\mu(y)}{|x-y|^{N-2}}$, we have

COROLLARY 15.4. *The Schwarz potential u_Γ of $\partial\Omega$ extends harmonically inside $\Omega \setminus \partial E$.*

The following quadrature identity is worth mentioning.

EXAMPLE 15.1. Let $\Omega = \left\{x \in \mathbb{R}^3 : x_2^2 + x_3^2 \le 1\right\}$ be the unit circular cylinder. Then for all harmonic functions u integrable in Ω the following quadrature formula holds

$$\int_\Omega u(x)\,dx = \pi \int_{-\infty}^{+\infty} u\left(x_1, 0, 0\right) dx_1. \tag{15.5}$$

We leave it as an exercise to check that the restriction of all such u on the x_1-axis are integrable so the right-hand side of (15.5) is well defined.

Now applying Corollary 15.2 adopted to the family of prolate ellipsoids

$$\Omega_a = \left\{\frac{x_1^2}{a^2} + x_2^2 + x_3^2 \le 1\right\},$$

we have

$$\begin{aligned}\int_\Omega u(x)\,dx &= \lim_{a\to\infty} \int_{\Omega_a} u(x)\,dx \\ &= \lim_{a\to\infty} \pi a\left(a^2-1\right)^{-3/2} \int_{-(a^2-1)^{1/2}}^{(a^2-1)^{1/2}} u\left(x_1, 0, 0\right)\left(a^2 - 1 - x_1^2\right) dx_1.\end{aligned}$$

Since $a\left(a^2-1\right)^{-3/2}\left(a^2-1-x_1^2\right)$ tends to 1 as $a \to \infty$, Lebesgue's dominated convergence theorem yields (15.5) (cf. [**110**, p. 74 ff.] for more details).

Notation. Set $q_{a(\lambda)}(x) := \sum_{j=1}^{N} \frac{x_j^2}{a_j^2+\lambda} - 1$.

The proof of MacLaurin's theorem yields the following

COROLLARY 15.5. *For any $m \in \mathbb{N}$ and any, say, entire harmonic function u the following "mean value property" holds*

$$\frac{1}{|\Omega_\lambda|} \int_{\Omega_\lambda} q_{a(\lambda)}^m(x)\, u(x)\,dx = \ const. \tag{15.6}$$

for all $\lambda > -a_N^2$.

In particular, arguing as before, we obtain another corollary.

COROLLARY 15.6. *The exterior potential u_{Ω, q_a^m} with density $p = q_a^m$ extends harmonically to $\mathbb{R}^N \setminus E$, and moreover, as a multivalued function to $\mathbb{R}^N \setminus \partial E$.*

PROOF OF COROLLARY 15.5. Set $b_j = \left(a_j^2 + \lambda\right)^{1/2}$. As in the proof of Theorem 15.1, by changing variables $x_j = a_j y_j$, $x_j = b_j y_j$, $j = 1, \ldots, N$, we reduce the statement to showing that

$$\int_B \left(|y|^2 - 1\right)^m \left(\sum_{j=1}^N a_j \xi_j y_j\right)^k dy = \int_B \left(|y|^2 - 1\right)^m \left(\sum_{j=1}^N b_j \xi_j y_j\right)^k dy$$

for all $\xi \in \Gamma_0$ (the isotropic cone) and all k (B is the unit ball in $\mathbb{R}^N$).

Or, equivalently, recalling that

$$\sum_{j=1}^N \left((a_j\xi_j)^2 - (b_j\xi_j)^2\right) = 0,$$

we have to show that

$$P(t) := \int_B \left(|y|^2 - 1\right)^m \left(\sum_{j=1}^N t_j y_j\right)^k dy$$

depends on $\sum_{j=1}^N t_j^2$ only. But since $\left(|y|^2 - 1\right)$ is invariant under rotations, we can extend word-for-word the argument used in proving the last assertion in the proof of Theorem 15.1. □

REMARK 15.3. In fact, we can replace the density $q_{a(\lambda}^m(x)$ by $f \circ q_{a(\lambda)}(x)$ where f is a more or less arbitrary analytic function of one variable. In particular, m in (15.6) need not be an integer.

Notes

Theorem 15.1 goes back to MacLaurin who considered prolate spheroids. The general case was treated later by Laplace (see [**139**] and the discussion therein; cf. [**106**]). The proof presented here is from [**110**]. A proof of the Asgeirsson theorem and the derivation of Theorem 15.1 from it can be found in [**26**, vol. II]. Fischer's product and its applications to partial differential operators with constant coefficients appeared in his seminal paper [**52**]. The reader is referred to [**147**, **148**] for more applications and other developments. The names of H. Weyl, V. Bargmann, V. A. Fock, and E. Bombieri have also been attached to this inner product, and in the theory of random polynomials the Gaussian measure induced by this inner product (each inner product naturally gives rise to a Gaussian probability distribution) is often referred to as the Kostlan-Shub-Smale model [**45**] [**177**]. Corollaries 15.2, 15.3, 15.4, and Example 15.1 are taken from [**110**].

CHAPTER 16

Potential Theory on Ellipsoids: Part II - There is No Gravity in the Cavity

Corollary 15.6 from the previous chapter revealed that the exterior Newtonian potential generated by uniform density on an ellipsoid extends harmonically into the region occupied by mass up to the so-called focal ellipsoid E. In the next section, we extend this result to the case of an arbitrary polynomial density.

Later in this chapter, after discussing single-layer potentials, we will prove Newton's theorem (and some interesting generalizations) on the vanishing of gravitational force inside the "cavity" of a uniformly massive ellipsoidal shell.

We keep the same notations as in Chapter 15.

1. Arbitrary polynomial density

THEOREM 16.1. *Let p be a polynomial. Then the potential*

$$u_{p,\Omega}(x) = c_N \int_\Omega \frac{p(y)}{|x-y|^{N-2}}\,dy \tag{16.1}$$

extends harmonically across $\Gamma := \partial\Omega$ into $\Omega \setminus E$.

PROOF. Observe that in view of Corollary 15.6 the theorem follows immediately from the following two lemmas. □

LEMMA 16.1. *Let Q denote the linear space generated by*

$$\left\{ \sum_{j=0}^{m'} \sum_{|\alpha|\le m-2j} A_\alpha \partial^\alpha q_a^{m-j} \right\}_{m=0}^{\infty},$$

where

$$m' = \begin{cases} \frac{m}{2}, & m \text{ even}, \\ \frac{m+1}{2}, & m \text{ odd} \end{cases}$$

$(q = q_a = \sum_1^N \frac{x_j^2}{a_j^2} - 1)$. *Then*

$$Q = P := \{\textit{all polynomials in } N \textit{ variables}\}.$$

LEMMA 16.2. *For any $m \in \mathbb{N}$, $\alpha : |\alpha| \le m$, $x \in \mathbb{R}^N \setminus \bar{\Omega}$,*

$$\int_\Omega \partial_y^\alpha q_a^m(y)\,|x-y|^{2-N}\,dy = \partial_x^\alpha \int_\Omega q_a^m(y)\,|x-y|^{2-N}\,dy.$$

PROOF OF LEMMA 16.1. The proof is by induction. Obviously $P_0 \subset Q$. Assume that $P_k \subset Q$. We want to show: $x_j x^\alpha$, $|\alpha| = k$, $j = 1, \ldots, N$ are all in Q.

Without loss of generality we may assume $j = 1$. By the induction hypothesis we have

$$x_1 x^\alpha = x_1 \sum_{j=0}^{k} \left(\sum_{|\alpha| \le k-2j} A_j \partial^\alpha q^{k-j}(x) \right)$$
$$= \sum_{j=0}^{k} \left(\sum_{|\alpha| \le k-2j} A_j x_1 \partial^\alpha q^{k-j}(x) \right).$$

So we only have to check that $x_1 \partial^\alpha q^{k-j} \in Q$. Note that for $\alpha = (\alpha_1, \dots, \alpha_n)$ with $\alpha_1 = 0$ there is nothing to prove. Hence, assume $\alpha_1 \ge 1$, and set $\beta = (\alpha_1 + 1, \alpha_2, \dots, \alpha_N)$ and $\sigma = (\alpha_1 - 1, \alpha_2, \dots, \alpha_N)$.

Now

$$x_1 \partial^\alpha q^{k-j} = \frac{a_1^2}{2(k+1-j)} \partial^\beta q^{k+1-j} - \alpha_1 \partial^\sigma q^{k-j}.$$

Indeed, if we write $\alpha = (\alpha_1, 0, \dots, 0) + \alpha'$, we have

$$\frac{a_1^2}{2(k+1-j)} \partial^\alpha \partial_1 q^{k+1-j} = \partial^{\alpha'} \left\{ \partial^{\alpha_1 - 1} \left(q^{k-j} + x_1 \partial_1 q^{k-j} \right) \right\}$$
$$= \partial^{\alpha'} \left\{ \partial^{\alpha_1 - 2} \left(2\partial_1 q^{k-j} + x_1 \partial_1^2 q^{k-j} \right) \right\}$$
$$= \cdots$$
$$= \alpha_1 \partial^\sigma q^{k-j} + x_1 \partial^\alpha q^{k-j}.$$

□

PROOF OF LEMMA 16.2. For $m = 0$ or $\alpha = 0$, there is nothing to prove. So, let $m > 0$. Fix $\alpha : |\alpha| \le m$ with, say, $\alpha_1 > 0$. Let $\sigma = (\alpha_1 - 1, \alpha_2, \dots, \alpha_N)$. Note that $\partial^\sigma q^m = 0$ on $\Gamma := \partial\Omega$. Therefore, integration by parts yields

$$\int_\Omega \partial_y^\alpha q^m(y) \, |x-y|^{2-N} \, dy = \int_\Omega \partial_1 \left(\partial^\sigma q^m(y) \right) |x-y|^{2-N} \, dy$$
$$= \int_\Omega \left\{ \partial_1 \left(\partial^\sigma q^m(y) \, |x-y|^{2-N} \right) \right\} dy - \int_\Omega \left(\partial^\sigma q^m(y) \right) \partial_1 \left\{ |x-y|^{2-N} \right\} dy$$
$$= \int_\Omega \left(\partial^\sigma q^m \right) \partial_{x1} \left\{ |x-y|^{2-N} \right\} dy = \partial_{x_1} \int_\Omega \left(\partial^\sigma q^m \right) |x-y|^{2-N} \, dy.$$

Iterating this argument we move the differential operator ∂^α outside the integral. This completes the proof of Lemma 16.2 and, hence, of Theorem 16.1. □

From Theorem 16.1 and Corollaries 15.5 and 15.6 it follows

COROLLARY 16.1. *The exterior potential (16.1) with a polynomial density can be represented as a finite linear combination of potentials of masses supported on the focal ellipsoid*

$$E := \left\{ x \in \mathbb{R}^N : x_N = 0, \sum_{1}^{N-1} \frac{x_j^2}{a_j^2 - a_N^2} - 1 \le 0 \right\}$$

with real-analytic densities and their derivatives. In particular, all such potentials $u_{p,\Omega}$ *extend as (multi-valued) harmonic functions to* $\mathbb{R}^N \setminus \partial E$.

2. The standard single layer potential

In general, if Γ is a compact, smooth, real hypersurface a single layer potential with density p is defined by

$$u(x) := c_N \int_\Gamma \frac{p(y)}{|x-y|^{N-2}}\, dS_y,$$

where dS_y is Lebesgue measure on Γ.

Clearly, u is harmonic in $\mathbb{R}^N \setminus \Gamma$ and continuous in $\mathbb{R}^N$. Consider the homothetic ellipsoids Γ_ε defined by

$$\Gamma_\varepsilon := \left\{ x \in \mathbb{R}^N : \sum_1^N \frac{x_j^2}{a_j^2} = 1 + \varepsilon \right\}.$$

Let Ω_ε be the "cavity" between Γ and Γ_ε. Take the potential $u_\varepsilon(x)$ of the uniform distribution on Ω_ε with density $1/\varepsilon$, then letting $\varepsilon \to 0$ we obtain

$$u_\varepsilon(x) \to u(x) = c_N \int_\Gamma \frac{\omega(y)}{|x-y|^{N-2}}\, dS_y,$$

where $u(x)$ is a single layer potential on Γ with density $\omega(y) = \frac{1}{|\nabla q_a(y)|}$, $y \in \Gamma$. We shall call $u(x)$ the standard single layer potential.

THEOREM 16.2. *Let p be a polynomial of degree m. Then, the potential*

$$u(x) = u_p(x) := c_N \int_\Gamma \frac{p\,\omega}{|x-y|^{N-2}}\, dA(y)$$

equals a harmonic polynomial of degree $\le m$ inside Ω.

COROLLARY 16.2 ("no gravity in the cavity"). *Let $t > 1$, $\tilde{\Omega} := t\Omega \setminus \Omega$ be an ellipsoidal cavity and let p be as above. Then, the volume potential with density p*

$$u_{p,\tilde{\Omega}}(x) := c_N \int_{\tilde{\Omega}} \frac{p}{|x-y|^{N-2}}\, dy$$

*equals a harmonic polynomial of degree $\le m$ inside Ω. In particular, for $p = const$, $u_{p,\tilde{\Omega}}(x) = const$ in Ω, i.e., $\nabla u_{p,\Omega} = 0$ and we obtain the theorem of Newton (cf. [**99**]): there is no attraction inside the ellipsoidal cavity.*

PROOF OF COROLLARY 16.2. In view of Theorem 16.2 and Fubini's theorem we have for $x \in \Omega$,

$$u_{p,\tilde{\Omega}}(x) = c_N \int_1^t \left\{ \int_{\Gamma_\tau} \frac{p\omega}{|x-y|^{N-2}}\, dS_\tau(y) \right\} d\tau = \int_1^t \left(\sum_{|\alpha| \le m} a_\alpha(\tau)\, x^\alpha \right) d\tau$$

which is a polynomial of degree $\le m$. □

PROOF OF THEOREM 16.2. For the sake of clarity we shall carry out the argument in $N = 3$ dimensions. The reader is encouraged to check that the same argument applies in all dimensions N. Fix $x_0 \in \Omega$ and a line ℓ through x_0. Let $\ell \cap \Gamma = \{x_1, x_2\}$. Take an infinitesimal circular cone of solid angle $d\theta$ with vertex x_0 and axis ℓ. It cuts Γ at two infinitesimal pieces S_1, S_2 and we are going to calculate the attraction at x_0 by S_1, S_2. Each piece S_i attracts x_0 with the force

$$f_i = \frac{m_i}{r_i^2}, \quad i = 1, 2, \tag{16.2}$$

where r_i are distances from S_i to x_0 (along ℓ),

$$m_i = \frac{p(x_i)\, dA_i}{|\nabla q_a(x_i)|} \tag{16.3}$$

are the masses of S_i and finally, the areas dA_i of S_i equal

$$dA_i = \frac{r_i^2 d\theta}{\cos\beta_i}, \quad i = 1, 2, \tag{16.4}$$

where β_i is the angle between the positive direction of ℓ (choose it to be $\overrightarrow{x_0 x_1}$) and the outer normal to Γ at x_i. Let t denote the parameter along the line ℓ. Without ambiguity we denote the restriction of q_a on ℓ by $q_a(t)$. Then (t_i denotes the value of the parameter corresponding to the point x_i, $i = 1, 2$), we have

$$\left.\frac{dq_a}{dt}\right|_{t_i} = |\nabla q_a(x_i)| \cos\beta_i. \tag{16.5}$$

Combining (16.2–16.5) and dividing by $d\theta$ we see that the attraction at x_0 by the infinitesimal elements S_i, $i = 1, 2$ along line ℓ equals

$$\sum_{i=1}^{2} \frac{p(t_i)}{q_a'(t_i)}. \tag{16.6}$$

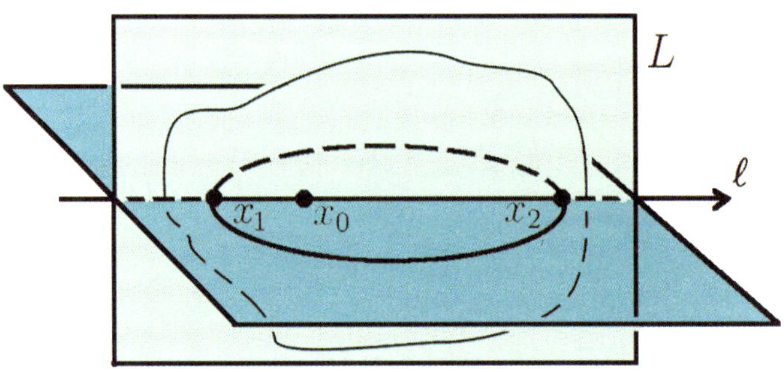

FIGURE 1. The force from two points is realized as a sum of residues in the complex line L.

Now consider the imbedding of ℓ into a complex line L (see Fig. 1), i.e., simply allow the parameter t to take on complex values. Then (16.6) is nothing else but

$$\sum \operatorname{res} \frac{p(t)}{q_a(t)}, \tag{16.7}$$

the sum of the residues of the rational function p/q_a on L. Expression (16.6) depends on point x_0 and a complex line L as parameters. Since calculating (16.7) reduces to a contour integral, we can interchange the differentiation of (16.6) with respect to x_0 and the operation of summing up the residues. Hence for any multi-index α we have

$$\partial_{x_0}^{\alpha}\left(\sum \operatorname{res} \left.\frac{p}{q_a}\right|_L\right) = \sum \left(\operatorname{res} \partial_{x_0}^{\alpha} \left.\frac{p}{q_a}\right|_L\right).$$

For $\alpha : |\alpha| \geq m :=$ degree of p, we observe that rational functions $\partial_{x_0}^{\alpha}(p/q)\big|_L$ are holomorphic at infinity and have a zero of order at least two there, as the following exercise shows.

EXERCISE 16.1. Prove that for $|\alpha| \geq m :=$ degree of p

$$\partial_{x_0}^{\alpha}\left(\frac{p}{q_a}\bigg|_L\right) = O\left(\frac{1}{t^2}\right), \qquad \text{as } t \to \infty \text{ in } L.$$

Now moving the contour of integration when calculating (16.7) to ∞ we obtain that for all $\alpha : |\alpha| \geq m$

$$\partial_{x_0}^{\alpha}\left[\sum \operatorname{res}\left(\frac{p}{q_a}\bigg|_L\right)\right] = 0. \tag{16.8}$$

Therefore, all derivatives of order $\geq m$ of the attraction exhibited on x_0 by infinitesimal elements of $\ell \cap \Gamma$ vanish. Summing (16.8) over all lines ℓ (or, strictly speaking, integrating (16.8) over $\mathbb{RP}^2$ or $\mathbb{RP}^{N-1}$ in general) we conclude that all derivatives of order $\alpha : |\alpha| \geq m$ of the attractive force $\nabla u_p(x_0)$ vanish. Hence, $u_p(x)$ is a polynomial of degree $\leq m$. The proof is now complete. □

3. Domains of hyperbolicity

In order to extend the above argument to a more general setting let us introduce the following definition.

DEFINITION 16.1. Let $\Gamma := \{x \in \mathbb{R}^N : \varphi(x) = 0, \varphi \in P_k\}$ be a smooth real algebraic surface of degree k with φ irreducible. A domain Ω is called a *domain of hyperbolicity* for Γ if for any $x_0 \in \Omega$ any line ℓ passing through x_0 intersects Γ at precisely k points.

For example, the interior of an ellipsoid is a domain of hyperbolicity. More generally, if a hypersurface of degree $2k$ consists of an increasing family of k ovaloids then the smallest one is the domain of hyperbolicity.

The standard single layer density on Γ is defined exactly as before with one distinction, the sign $+$ or $-$ is assigned on each connected component of Γ depending whether the number of obstructions for "viewing" this component from the domain of hyperbolicity of Γ is even or odd.

Then the above proof extends word-for-word to yield the following generalization of Newton's theorem:

THEOREM 16.3. *Let $\Gamma := \{x \in \mathbb{R}^N : \varphi(x) = 0, \varphi \in P_k\}$ be an algebraic surface of degree k and Ω its domain of hyperbolicity. Then for any polynomial $p \in P_m$ the single layer potential of the density $p\omega$*

$$u_p(x) := c_N \int_{\Gamma} \frac{p\omega}{|x-y|^{N-2}}\, dS(y),$$

where ω is the standard single layer density equals a harmonic polynomial of degree $\leq m - k + 2$ inside Ω. So, in particular, for $m \leq k - 2$ there is no attraction inside the domain of hyperbolicity.

4. The Schwarz potential conjecture for ellipsoids

Returning to ellipsoidal surfaces, we first notice the following corollary of Corollary 16.2.

COROLLARY 16.3. *If Ω is an ellipsoid and $p \in P_m$ then inside Ω the volume potential*

$$u_{p,\Omega} := c_N \int_\Omega \frac{p(y)}{|x-y|^{N-2}}\, dy$$

is a harmonic polynomial of degree $\le m+2$.

PROOF OF COROLLARY 16.3. For $t > 1$, let $\Omega_t := \left\{x \in \mathbb{R}^N : \sum_1^N \frac{x_j^2}{a_j^2} \le t^2\right\}$ denote the ellipsoid homothetic with Ω. Without loss of generality we can assume p to be homogeneous, i.e., $p \in H_m$. Fix $x \in \Omega$. Using homogeneity and a simple change of variables we calculate

$$u_{p,\Omega_t}(x) := c_N \int_{\Omega_t} \frac{p(y)}{|x-y|^{N-2}}\, dy = t^{m+2} u_{p,\Omega}\left(\frac{x}{t}\right). \tag{16.9}$$

By Corollary 16.2 the potential of the cavity $\Omega_t \setminus \Omega$ inside Ω,

$$u_{p,\Omega_t}(x) - u_{p,\Omega}(x) =: h_m(x) \in P_m, \tag{16.10}$$

is a harmonic polynomial. For any $\alpha : |\alpha| = m+2$ using (16.9–16.10) we obtain

$$\partial^\alpha u_{p,\Omega_t}(x) = \partial^\alpha \left[t^{m+2} u_{p,\Omega}\left(\frac{x}{t}\right)\right] = \partial^\alpha u_{p,\Omega}\left(\frac{x}{t}\right) = \partial^\alpha u_{p,\Omega}(x),$$

i.e., all such $\partial^\alpha u_{p,\Omega}$ are homogeneous and of degree 0 in Ω. Also, for any $\alpha : |\alpha| > m$ we have, since $\deg p = m$,

$$\Delta\left(\partial^\alpha u_{p,\Omega}(x)\right) = \partial^\alpha\left(\Delta u_{p,\Omega}(x)\right) = \partial^\alpha(-p) = 0.$$

Thus for all $\alpha : |\alpha| = m+2$, $\partial^\alpha u_{p,\Omega}(x)$ are harmonic and homogeneous of degree 0. Hence, all $\partial^\alpha u_{p,\Omega}(x)$ are constants for $\alpha : |\alpha| = m+2$ and the Corollary follows. □

REMARK 16.1. (i) In particular, the potential of the uniform density is a quadratic polynomial inside the ellipsoid. This is an old theorem of Newton (cf. [**99**, **139**]).

(ii) It turns out that even a partial converse to Corollary 16.3 is also true. Namely, if a potential of the uniform density in a solid K equals a quadratic polynomial inside K, K must be an ellipsoid. In dimensions 2 and 3 this was proved, respectively, by Hölder [**84**] and, independently, by P. Dive and W. Nikliborc [**34**, **149**]; cf. [**106**]. In full generality the result has been obtained by E. DiBenedetto and A. Friedman [**32**].

By translation and rotation one easily sees that the potential $u_\Omega(x)$ can be written as $u_\Omega(x) := B - \sum\limits_1^N A_j x_j^2$ inside ellipsoid Ω, where $B > 0$, $A_j > 0$, $\sum\limits_1^N A_j = \frac{1}{2}$. The main difficulty in proving the DiBenedetto-Friedman theorem is to show that in fact every such polynomial represents an interior potential of some ellipsoid. Recently a simple proof of the latter assertion was obtained by L. Karp [**97**] by making use of L. Brouwer's domain invariance theorem. Assuming this, the following beautiful argument of Dive finishes the proof. Namely, find an ellipsoid Ω whose interior potential is the same quadratic polynomial. By Corollary 16.2 all ellipsoids homothetic with Ω, have interior potentials equal $B' - \sum_1^N A_j x_j^2$ with a different constant term, but the same coefficients A_j. Let $t_0 = \inf\{t : \Omega_t \supset K\}$. Then there exists $x_0 \in \partial\Omega_{t_0}$ such that $x_0 \in \partial K$ and Ω_{t_0} and K have a common supporting plane H at x_0. The potential of the shell between Ω_{t_0} and K (i.e., $\Omega_{t_0} \setminus K$) is constant inside K. Thus the attractive force vanishes identically inside K. In particular, the attraction in the direction of the normal to H at x_0 vanishes. This leads to an obvious contradiction unless Volume$(\Omega_{t_0} \setminus K) = 0$, i.e., $\Omega_{t_0} = K$.

(iii) It is easy to compute an equilibrium potential of an ellipsoid Ω. Recall (cf. [**99**]) that $V(x)$ is called an equilibrium potential of a total charge σ on $\Gamma := \partial\Omega$ if

$$V(x) := c_N \int_\Gamma \frac{\rho(y)}{|x-y|^{N-2}}\, dA(y),$$

where $\rho(y)$ is the density function, so that $V(x) \equiv 1$ inside Ω and

$$\lim_{|x|\to\infty} |x|^{N-2} V(x) = \sigma.$$

Then we have the following (as above, the volume potential $u_\Omega(x) = B - \sum\limits_1^N A_j x_j^2$ inside Ω).

COROLLARY 16.4.

$$V(x) = \frac{1}{B}\left(\hat{\mu}(x) - \frac{1}{2}\sum_1^N x_j \frac{\partial \hat{\mu}}{\partial x_j}(x)\right), \tag{16.11}$$

where (cf. Corollary 15.2) E is the focal ellipsoid,

$$\hat{\mu}(x) := c_N \int_E \frac{d\mu(y')}{|x-y|^{N-2}}$$

and

$$d\mu(y') = 2\left(\frac{\prod\limits_{j=1}^N a_j}{\prod\limits_1^{N-1}\left(a_j^2 - a_N^2\right)^{1/2}}\right)\left(1 - \sum_1^{N-1} \frac{y_j^2}{a_j^2 - a_N^2}\right)^{1/2} dy'\big|_E.$$

In particular, the equilibrium potential extends harmonically to $\mathbb{R}^N \setminus \partial E$.

PROOF. The right side of (16.11) is harmonic in $\mathbb{R}^N \setminus \bar{\Omega}$, vanishes at ∞ and since the exterior potential $u_\Omega(x)$ equals (Corollaries 16.3 and 15.2) $B - \sum\limits_1^N A_j x_j^2$ inside Ω, and $\hat{\mu}(x)$ outside Ω (and is C^1-smooth in $\mathbb{R}^N$), we obtain that it equals 1 on $\Gamma := \partial\Omega$. Hence, it must coincide with the equilibrium potential $V(x)$ outside Ω. □

Let us mention in passing that the classical theorem of Ivory (cf., e.g., [**139**]) states that the equipotential surfaces of V are precisely ellipsoids confocal with $\Gamma := \partial\Omega$. To derive this result from Corollary 16.4, one simply notes that the right hand side of (16.11) changes only by a constant factor when Ω is replaced by a confocal ellipsoid

$$\Omega_\lambda := \left\{ x : \sum^N \frac{x_j^2}{a_j^2 + \lambda} \leq 1, \lambda \geq 0 \right\}.$$

Namely, $B \to B_\lambda$ while $\frac{d\mu_\lambda}{d\mu} = \frac{\mathrm{Vol}(\Omega_\lambda)}{\mathrm{Vol}(\Omega)}$.

For the classical proof of Ivory's Theorem, we refer the reader to [**55**, Lecture 30].

Finally, we are ready to prove the Schwarz Potential Conjecture for polynomial data on ellipsoidal surfaces.

COROLLARY 16.5. *All solutions u of Cauchy problems*

$$\begin{cases} \Delta u = 0, \\ u \equiv p \text{ on } \Gamma, \end{cases}$$

where data p is a polynomial extend harmonically to $\mathbb{R}^N \setminus \partial E$.

PROOF. Consider the potential

$$v(x) := c_N \int_\Omega \frac{(-\Delta p)(y)}{|x-y|^{N-2}}.$$

By Corollary 16.3 $v(x)\big|_\Omega = p + h$, where h is a harmonic polynomial and $\deg(h) \leq \deg(p)$. Also, Theorem 16.1 implies that $v(x)$ extends harmonically to $\mathbb{R}^N \setminus \partial E$. Call this extension $v_1(x)$. But on $\Gamma := \partial\Omega$ we have

$$u \equiv p \equiv v - h \equiv v_1 - h.$$

Hence $u := v_1 - h$ provides the desired extension. □

Notes

Lemmas 16.1 and 16.2 are from H. Shahgholian's paper [**172**], as well as Theorem 16.1. However, from there he follows a different route to Corollaries 16.3 and 16.5 elaborating on a beautiful idea of Dirichlet. Theorem 16.2 is due to M.N. Ferrers [**51**] for dimension $N = 3$. The elegant proof here and, what is more important, the far reaching extension (Theorem 16.3) are due to A. Givental [**65**]. Yet the crucial step of introducing the notion of a domain of hyperbolicity (Definition 16.1) is due to V.I. Arnold [**9**] as well as the partial case (although, by a different argument) of Theorem 16.3 when $m \leq k - 2$ which is a direct extension of Newton's theorem (cf. Corollary 16.2). For further interesting developments we refer the reader to a paper by B. Shapiro and A. Vainstein [**173**]. Corollary 16.3 is also due to Ferrers [**51**] in the case $N = 3$, although his proof, based on heavy computations with ellipsoidal harmonics, is hardly extendable to higher dimensions. Corollary 16.4 has been observed in [**110**]. Far reaching generalizations of Corollary 16.5 with arbitrary entire data and for more general differential operators with lower order terms have been obtained by G. Johnsson [**94**] and, independently and by different methods, by V. Shatalov and B. Sternin [**181**]. Domains of hyperbolicity sometimes coincide with so-called "spectrahedra" [**197**] which play an important role in convex optimization where they appear as constraint sets for interior-point methods [**79, 146**]. A spectrahedron is a region (in the parameter space of coefficients) of positive-definiteness for a linear combination of given matrices. The correspondence between domains of hyperbolicity and spectrahedra in the three-dimensional case is equivalent to a 1958 conjecture of P. Lax [**122**] that was proved by A. Lewis, P. Parrilo, and M. Ramana [**130**] using a groundbreaking result of J.W. Helton and V. Vinnikov [**80**].

CHAPTER 17

Potential Theory on Ellipsoids: Part III - The Dirichlet Problem

Given a homogeneous polynomial p of degree d in n variables, the Almansi expansion (see [**10**, **180**] and Exercise 17.1 below) expresses $p(x)$ as a sum

$$p(x) = h_d(x) + |x|^2 h_{d-2}(x) + |x|^4 h_{d-4}(x) + ..., \tag{17.1}$$

where each h_{d-2j} is a harmonic homogeneous polynomial of degree $d - 2j$. In particular, we notice (by taking $|x|^2 = 1$) that p matches a harmonic polynomial on the unit sphere. Since one may apply this to the homogeneous parts of a general polynomial, we immediately observe the well-known fact that the Dirichlet problem for Laplace's equation in the ball with polynomial boundary data always has a polynomial solution. It turns out that this algebraic aspect of the Dirichlet problem can be pursued further, and in the next sections we consider the case of ellipsoidal domains. That is, letting

$$\Omega = \left\{ x \in \mathbb{R}^n : \sum_{j=1}^{n} \frac{x_j^2}{a_j^2} - 1 \leq 0, \quad a_1 \geq a_2 \geq .. \geq a_n > 0 \right\}$$

be an ellipsoid, we consider the problem in Ω of finding a function $u \in C^2(\Omega) \bigcap C(\overline{\Omega})$ satisfying

$$\begin{cases} \Delta u = 0 \quad \text{in } \Omega, \\ u\big|_{\partial\Omega} = f, \end{cases} \tag{17.2}$$

where $\Delta = \sum_{j=1}^{n} \frac{\partial^2}{\partial x_j^2}$.

EXERCISE 17.1. Prove that every homogeneous polynomial p of degree d admits an Almansi expansion as stated in (17.1). Hint: Apply Lemma 14.1 recursively.

Note that one can deduce more directly the above conclusion about the Dirichlet problem in a ball from Lemma 14.1.

1. The Dirichlet problem in an ellipsoid: Polynomial data

PROPOSITION 17.1. *Suppose the data function $f = p$ is a polynomial of n variables. Then the solution u to the Dirichlet problem (17.2) is a harmonic polynomial. Moreover,*

$$\deg u \leq \deg p. \tag{17.3}$$

The proof, which is essentially linear algebra, is very similar to that of Lemma 14.1.

PROOF. Denote by $P_{n,m} = P_m$ the finite-dimensional space of polynomials of degree $\leq m$ in n variables. Let $q(x) = \sum \frac{x_j^2}{a_j^2} - 1$ be the defining quadratic for $\partial\Omega$. Consider the linear operator $T : P_m \to P_m$ defined by

$$T(r) := \Delta(qr).$$

The maximum principle implies that $\ker T = 0$, so T is injective. Since $\dim P_m < \infty$, this implies that T is surjective.

Hence, given $P \in P_m$ with $m \geq 2$, we can find a polynomial $r \in P_{m-2}$ such that $T(r) = \Delta P$. The function

$$u = P - qr$$

is then the solution of (17.2). □

2. Entire data

The following extension to the case of entire data is due to H.S. Shapiro and the first author [**113**]; cf. [**175**]. The proof we give here is adapted from [**113**] (with minor changes). The reader will find several main ideas in the proof already present in the discussion of the CP problem for the spheres in Ch. 14.

THEOREM 17.1 ([**113**]). *Suppose the data function f extends as an entire function to $\mathbb{C}^n$. Then the solution u of the Dirichlet problem (17.2) extends to a harmonic function on $\mathbb{R}^n$. (Hence u extends to an entire function on $\mathbb{C}^n$.)*

We will use the following basic criterion for entireness; see [**155**, Prop. 12].

PROPOSITION 17.2 (cf. Lemma 14.3). *Let $f = \sum_{m=0}^{\infty} f_m$, where f_m is a homogeneous polynomial of degree m. Then f is entire if and only if*

$$\lim_{m\to\infty} \left(\max_{\theta \in S^{n-1}} |f_m(\theta)| \right)^{1/m} = 0. \tag{17.4}$$

PROOF. Suppose f extends to an entire function on $\mathbb{C}^n$. Then $f_m(z)$ are the Taylor coefficients of an entire function $t \mapsto f(tz)$ defined on $\mathbb{C}$, and the Cauchy-Hadamard estimate gives, for each $T > 1$,

$$|f_m(z)| \leq \frac{\max\{|f(tz)| : |t| = T\}}{T^m},$$

and hence

$$\max_{|z|=1} |f_m(z)| \leq \frac{\max\{|f(\zeta)| : |\zeta| = T\}}{T^m}.$$

Taking mth roots and letting $m \to \infty$ gives

$$\limsup_{m\to\infty} \left(\max_{|z|=1} |f_m(z)| \right)^{1/m} \leq T^{-1}$$

for arbitrary $T > 1$, implying (17.4). To see the converse, we assume (17.4) and use the fact that since f_m is homogeneous, we have (see [**178**])

$$\max_{|z|=1} |f_m(z)| \leq 2^{m/2} \max_{\theta \in S^{n-1}} |f_m(\theta)|. \tag{17.5}$$

Together with (17.4) this implies that $M_m := \max_{|z|=1} |f_m(z)|$ satisfies $M_m^{1/m} \to 0$ as $m \to \infty$. Thus, $\sum_{m=0}^{\infty} |f_m|$ is locally uniformly convergent on $\mathbb{C}^n$, which implies that f is entire. □

EXERCISE 17.2. Use the following steps to prove the estimate (17.5) in the case $n = 2$ (cf. [**10**, p. 55-62]). That is, for $f_m(X, Y)$ a homogeneous polynomial of degree m, we have

$$\max_{(X,Y)\in\mathbb{C}^2:|X|^2+|Y|^2=1} |f_m(X,Y)| \leq 2^{m/2}\left(\max_{(x,y)\in\mathbb{R}^2:x^2+y^2=1} |f_m(x,y)|\right). \tag{17.6}$$

(a) Show that the unit ball $\{(X,Y) \in \mathbb{C}^2 : |X|^2 + |Y|^2 < 1\}$ is contained in the bidisk $B := \{(z,w) \in \mathbb{C}^2 : |z| < \sqrt{2}, |w| < \sqrt{2}\}$, defined in terms of the variables $z = X + iY$, $w = X - iY$.
(b) Show that the polynomial $p(z,w) = f_m(X+iY, X-iY)$ attains its maximum modulus over B at a point (z_0, w_0) with $|z_0| = 1$ and $|w_0| = 1$. (This result is a special case of a more general statement about the "Shilov boundary" of a polydisk.) Hint: Fixing one variable, apply the (univariate) maximum modulus principle in the other variable.
(c) Use homogeneity of p (which follows from homogeneity of f_m) to compare the maximum of $|p|$ over $|z| = \sqrt{2}, |w| = \sqrt{2}$ to that over $|z| = 1$, $|w| = 1$.
(d) Choose θ so that multiplication of (z_0, w_0) by the complex scalar $e^{i\theta}$ results in a point on the set where $w = \bar{z}$. Use homogeneity of p to see that $|p|$ takes the same value at the new point. Recall how this relates to the original variables X, Y in order to derive (17.6).

PROOF OF THEOREM 17.1. Without loss of generality (making a dilation if necessary), we may assume that Ω contains the unit ball. We also take $B = \{|x| \leq \rho\}$ to be a ball that contains Ω (say choosing the radius $\rho = a_1$ to be the largest semi-axis of Ω). Write the Taylor expansion of f as $f = \sum_{m=0}^{\infty} f_m$, as a sum of homogeneous polynomials f_m. Then by Proposition 17.2 we may write

$$\max_{\theta\in S^{n-1}} |f_m(\theta)| = \varepsilon_m^m$$

with $\varepsilon_m \to 0$ as $m \to \infty$. Let $u_m \in P_m$ be the solution, guaranteed by Proposition 17.1, to the Dirichlet problem with boundary data f_m, and let

$$u_m = u_{m,0} + u_{m,1} + \dots + u_{m,m}$$

be its decomposition into homogeneous polynomials. We will show that the solution

$$u(x) = \sum_{m=0}^{\infty}\sum_{k=0}^{m} u_{m,k}(x)$$

to the Dirichlet problem with boundary data f is entire.

For any fixed j, the degree j homogeneous term of u is given by $\sum_{m=j}^{\infty} u_{m,j}$. Assume that each of these sums converges for $\theta \in S^{n-1}$ (this is justified throughout the estimates that follow). Then, according to Proposition 17.2, u is entire if

$$\limsup_{j\to\infty}\left(\max_{\theta\in S^{n-1}}\left|\sum_{m=j}^{\infty} u_{m,j}(\theta)\right|\right)^{1/j} = 0 \tag{17.7}$$

Let us first estimate $|u_{m,j}(\theta)|$ for $\theta \in S^{n-1}$:

Applying Lemma 14.4, we have, for some constant C_n depending only on n:

$$\begin{aligned}
\max_{\theta\in S^{n-1}} |u_{m,j}(\theta)| &\le C_n j^{n/2} \max_{\theta\in S^{n-1}} |u_m(\theta)| \\
\text{(by the maximum principle)} \qquad &\le C_n j^{n/2} \max_{x\in\partial\Omega} |u_m(x)| \\
\text{(boundary data)} \qquad &= C_n j^{n/2} \max_{x\in\partial\Omega} |f_m(x)| \\
\text{(by homogeneity and } \Omega\subset B) \qquad &\le C_n j^{n/2} \max_{x\in\partial B} |f_m(x)| \\
\text{(by homogeneity)} \qquad &= C_n j^{n/2} \rho^m \max_{\theta\in S^{n-1}} |f_m(\theta)| \\
&= C_n j^{n/2} \rho^m \varepsilon_m^m.
\end{aligned}$$

Next, we show u is entire by establishing the condition (17.7). Starting with a simple estimate followed by term-wise application of the above, we obtain:

$$\begin{aligned}
\max_{\theta\in S^{n-1}} \left| \sum_{m=j}^{\infty} u_{m,j}(\theta) \right| &\le \sum_{m=j}^{\infty} \max_{\theta\in S^{n-1}} |u_{m,j}(\theta)| \\
&\le C_n j^{n/2} \sum_{m=j}^{\infty} \rho^m \varepsilon_m^m.
\end{aligned}$$

Taking the jth root, we notice that $\left(C_n j^{n/2}\right)^{1/j} \to 1$, as $j\to\infty$. Hence, taking $\delta_m = \rho\varepsilon_m$ in the following elementary exercise establishes the condition (17.7) and concludes the proof of Theorem 17.1. □

EXERCISE 17.3. Assume that δ_m is a sequence of numbers satisfying $(\delta_m)\to 0$ as $m\to\infty$. Then show that

$$\lim_{j\to\infty} \left(\sum_{m=j}^{\infty} \delta_m^m \right)^{1/j} = 0.$$

3. The Khavinson-Shapiro conjectures

In [**113**] the authors conjectured that the converse to Theorem 17.1 also holds, i.e., ellipsoids are the only domains in which all (DP) with entire data have entire solutions. More precisely, the following two statements are usually referred to as the Khavinson-Shapiro conjectures (cf. [**155**], [**156**], [**158**], [**137**]).

CONJECTURE 17.1 (KS1). *Ellipsoids are the only bounded domains for which polynomial data implies polynomial solution for the Dirichlet problem (*17.2*).*

CONJECTURE 17.2 (KS2). *Ellipsoids are the only bounded domains for which entire data implies entire solution for the Dirichlet problem (*17.2*).*

In other words, the conjectures state that Proposition 17.1 (without the degree condition (17.3)) and Theorem 17.1 each characterize ellipsoids.

In recent years, a substantial progress has been achieved by H. Render [**155**] (cf. the Notes at the end of this chapter and the next).

4. The Brelot-Choquet theorem and harmonic divisors

In the proof of Proposition 17.1 one may instead employ, in place of the maximum principle, the Brelot-Choquet theorem [**23**], an algebraic result that is extremely useful in the context of harmonic polynomials.

THEOREM 17.2 ([**23**]). *Suppose that h is a harmonic polynomial of degree $m > 1$ such that $h = h_1 \cdot h_2$ with h_1, h_2 each having degree $< m$. Then h_1 and h_2 must change sign.*

EXERCISE 17.4 (Proof of the Brelot-Choquet theorem). Use the following steps to give a proof of the Brelot-Choquet theorem.

(a) Argue that we may consider without loss of generality the case when h is homogeneous.
(b) Assume for the sake of contradiction that $h = h_1 \cdot h_2$ factors as a product with h_1, h_2 each homogeneous of degree $< m$ with, say, $h_1 \geq 0$.
(c) Use the fact (see Exercise 14.1–cf. [**180**]) that homogeneous harmonic polynomials of different degrees are orthogonal in the space $L^2(S^{n-1})$ in order to arrive at a contradiction.

EXERCISE 17.5. The following exercises revisit Proposition 17.1 from the point of view of the Brelot-Choquet theorem.

(a) Show that the Brelot-Choquet theorem can be used in place of the maximum principle in the step of the proof of Proposition 17.1 where injectivity was established.
(b) Based on this alternative approach, show that Proposition 17.1 generalizes to certain unbounded quadrics such as paraboloids and cylinders with ellipsoidal base.
(c) Show that Proposition 17.1 does not generalize to all quadrics by carefully choosing a specific hyperboloid for which injectivity fails. (For simplicity, take $n = 3$.)

Notes

Proposition 17.1 was widely known in the nineteenth century for $n = 2, 3$ (perhaps due to Lamé) and was proved with the use of ellipsoidal harmonics. The idea of the proof presented here goes back at least to Fischer [**52**].

The extension to entire functions in Theorem 17.1 was proved in [**113**]. This result was sharpened by Armitage in [**8**] who showed that the solution's order and type are dominated by that of the data.

One might ask whether the Proposition 17.1 extends to say rational or algebraic data, i.e., does having a smooth data function f that is a rational (algebraic) function of $x_1, x_2, .., x_n$ imply that the solution u is as well rational (algebraic)? The answer is a resounding "no" but the proofs become technically more involved; see [**17, 18, 37, 43**].

The two conjectures (KS1) and (KS2) will be discussed in Chapter 18, where, in particular, we state a result of H. Render, who has used "real Fischer spaces" to confirm both conjectures for a large class of algebraic surfaces [**155**].

The classical Dirichlet problem of finding a harmonic function a domain $\Omega \subset \mathbb{R}^n$ matching a given continuous function f on $\partial\Omega$ gained huge attention in the second half of the nineteenth century due to its central role in Riemann's proof of the existence of a conformal map of any simply connected domain onto the disk. Riemann's proof was later criticized by Weierstrass, and, after a considerable turmoil, corrected and completed by Hilbert and Fredholm; see [**154**] for a very nice historical account.

The Brelot-Choquet Theorem has an extensive follow up under the general umbrella of "harmonic divisors". The interested reader may start with H. Render's paper [**157**] and find further references therein.

CHAPTER 18

Singularities Encountered by the Analytic Continuation of Solutions to the Dirichlet Problem

Let us continue the discussion of the Dirichlet problem, which we restate here for the reader's convenience.

$$\begin{cases} \Delta u = 0, \\ u|_{\partial\Omega} = v. \end{cases} \tag{18.1}$$

We mostly focus on the case when Ω is homeomorphic to a ball with real-analytic, or even algebraic, boundary.

As the reader will see, the subject of the singularities to the solutions of the DP is not at all lacking in open problems, in spite of some progress achieved in recent decades.

Inspired by the direction taken in the previous chapter (and in particular, the result stated in Theorem 17.1), we ask the following broad question.

Question. Suppose the data v is a restriction to $\Gamma := \partial\Omega$ of a "very good" function, say an entire function of variables $x_1, x_2, ..., x_n$. In other words, the data presents no reasons whatsoever for the solution u of (18.1) to develop singularities.

(i) For which Γ can we then assert that all solutions u of (18.1) with entire data v are also entire?

(ii) If singularities do occur, they must be caused by geometry of the hypersurface Γ interacting with the differential operator Δ. Can we then find data v_0 that would force the worst possible scenario to occur? More precisely, for any entire data v, the set of possible singularities of the solution u of (18.1) is a subset of the singularity set of u_0, the solution of (18.1) with data v_0.

Remark 18.1. Addressing question (i), we know (by Theorem 17.1) that ellipsoids have the desired property. Under the assumption that Ω is bounded, if conjecture (KS2) is true, then there are no other examples.

It is well known since the early 20th century from works of Poincare, C. Neumann, Hilbert, and Fredholm that the solution u exists and is unique. Also, since u is harmonic in Ω, hence real-analytic there, no singularities can appear in Ω, and the solution u extends as an analytic function of n complex variables to the harmonicity hull (or Vekua hull–cf. Ch. 9, Sec. 2) $\widehat{\Omega} \subset \mathbb{C}^n$ of Ω. Moreover, assuming $\Gamma := \partial\Omega$ to consist of real-analytic hypersurfaces, the more recent and difficult results on "elliptic regularity" assure us that if the data v is real-analytic in a neighborhood of $\overline{\Omega}$ then u extends as a real-analytic function across $\partial\Omega$ into an open neighborhood Ω' of $\overline{\Omega}$. In two dimensions, this can be done using the reflection principle. In higher dimensions, the boundary can be biholomorphically "flattened", but this leads to a general elliptic operator for which the reflection principle does not apply. Instead, analyticity must be shown by directly verifying convergence of the power series representing the solution through difficult estimates on the derivatives (see [**53**]).

1. The Dirichlet problem: When does entire data imply entire solution?

In 2005 H. Render [**155**] proved the conjecture (KS2) stated in the previous chapter under the assumption that the senior homogeneous part $\phi_N(x)$ of ϕ is *elliptic*, i.e., $|\phi_N(x)| \geq C|x|^N$ for some constant $C > 0$. This entails an algebro-geometric restriction, namely, that Ω is a component of the sublevel set $\{\phi(x) < 0\}$ of a polynomial ϕ whose zero set $\{\phi(x) = 0\}$ is a *bounded* set in $\mathbb{R}^n$.[1] For $n = 2$, a simpler version of this result (for polynomial data) was observed in 2001 by M. Chamberland and D. Siegel [**25**]. We outline their argument in the next section.

REMARK 18.2. We will return to Render's theorem below. For now let us note that, unfortunately, it leaves open simple examples such as (in $n = 2$ dimensions) perturbations of a unit disk, e.g., where Ω is the connected component of the set $\{x \in \mathbb{R}^2 : x^2 + y^2 - 1 + \varepsilon x^3 < 0\}$ that contains the origin, which is still bounded if ε is taken sufficiently small.

2. When does polynomial data imply polynomial solution?

In this section we return to the simpler conjecture (KS1) regarding polynomial data.

Let $\gamma = \{\phi(x) = 0\}$ be a bounded, irreducible algebraic curve in $\mathbb{R}^2$. If the DP posed on γ has polynomial solution whenever the data is a polynomial, then as Chamberland and Siegel observed, (a) γ is an ellipse or (b) there exists data $f \in \mathbb{P}$ such that the solution $u \in \mathbb{P}$ of DP has $\deg u > \deg f$.

In case (b) $u - f\big|_\gamma = 0$ implies that ϕ divides $u - f$ by Hilbert's Nullstelensatz [**121**, Ch. IX, Sec. 1], and, since $\deg u = M > \deg f$, $u_M = \phi_k g_l$ where ϕ_k and u_M are the senior homogeneous terms of ϕ and u respectively. The senior term of u must have the form $u_M = az^M + b\bar{z}^M$ since u_M is harmonic. Hence, u_M factors into linear factors and so must ϕ_k. Hence γ is unbounded. This gives the following result [**25**].

THEOREM 18.1. *Suppose deg $\phi > 2$ and ϕ is square-free. If the Dirichlet problem posed on $\{\phi = 0\}$ has a polynomial solution for each polynomial data, then the senior part of ϕ, which we denote by ϕ_N, of order N, factors into real linear terms, namely,*

$$\phi_N = \prod_{j=0}^{N} (a_j x - b_j y),$$

where a_j, b_j are some real constants and the angles between the lines $a_j x - b_j y = 0$, for all j, are rational multiples of π.

This theorem settles Conjecture (KS1) for bounded planar domains

$$\Omega \subseteq \{\phi(x) < 0\}$$

such that the set $\{\phi(x) = 0\}$ is bounded in $\mathbb{R}^2$.

---

[1] Note that this is a stronger statement than merely assuming that $\partial\Omega$ is bounded since the level set $\{\phi = 0\}$ may have unbounded components in addition to $\partial\Omega$.

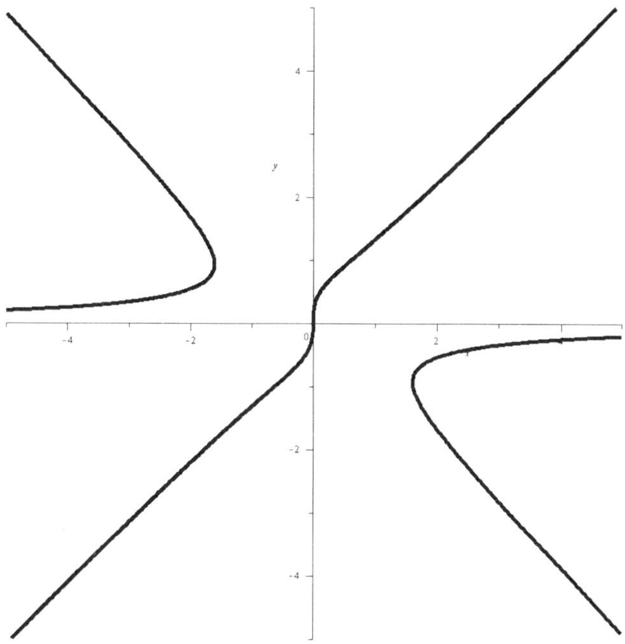

FIGURE 1. A plot of the cubic $y(y-x)(y+x)-x=0$. Any quadratic data can be matched by a harmonic polynomial on this set.

EXAMPLE 18.1. The curve $y(y-x)(y+x)-x=0$ (see Figure 1) satisfies the necessary condition imposed by the theorem. Moreover, any quadratic data can be matched on it by a harmonic polynomial. For instance, $u = xy(y^2 - x^2)$ solves the interpolation problem (it is misleading to say "Dirichlet" problem, since there is no bounded component) with data $v(x, y) = x^2$. On the other hand, the data x^3 does not have polynomial solution as the following exercise shows.

EXERCISE 18.1. Use the following steps to prove that the data x^3 does not match any harmonic polynomial on the curve defined by the zero set of $\psi(x, y) := y(y-x)(y+x)-x$.

(a) Assume for the sake of contradiction that there is such a harmonic polynomial $h(x, y)$ and apply Hilbert's Nullstelensatz as above to conclude that $h(x, y) - x^3$ factors as $\psi(x, y) \cdot q(x, y)$.
(b) Write this in terms of the homogeneous expansions of ψ, q, and h. Equating homogeneous parts, argue that $q_0 \neq 0$, $q_2 \neq 0$.
(c) Proceeding inductively, argue that $q_{2k} \neq 0$ for every k, which is a contradiction since q is a polynomial.

Note that for unbounded γ as in the above example, although the interpolation problem is not a genuine Dirichlet problem, it yet fits into the global Goursat problem discussed in [**175**]; cf. Section 4 of this Chapter.

3. The Dirichlet problem and Bergman orthogonal polynomials

In two dimensions, the conjecture (KS1) was confirmed under a degree-related condition on the solution in terms of the data [**116**]. This utilized a surprising equivalence, established by N. Stylianopoulos and M. Putinar [**152**], of the conjecture to the existence of finite-term recurrence relations for Bergman orthogonal polynomials. In order to state the degree conditions and the associated recurrence conditions, assume Ω is a domain in $\mathbb{R}^2$ with C^2-smooth boundary. Let $\{p_m(z)\}$ be the Bergman orthogonal polynomials (orthogonal with respect to area measure over Ω). These are analytic polynomials of the complex variable z. Consider the following properties for Ω.

(1) There exists C such that for a polynomial data of degree m there always exists a polynomial solution of the Dirichlet problem posed on Ω of degree $\leq m + C$.
(2) There exists N such that for all k, m, the solution of the Dirichlet problem with data $\bar{z}^k z^m$ is a harmonic polynomial of degree $\leq (N-1)k + m$ in z and of degree $\leq (N-1)m + k$ in $\bar{z}$.
(3) There exists N such that $\{p_m\}$ satisfy a (finite) $(N+1)$-recurrence relation, i.e., there are constants $a_{m-j,m}$ such that

$$zp_m = a_{m+1,m}p_{m+1} + a_{m,m}p_m + \ldots + a_{m-N+1,m}p_{m-N+1}.$$

(4) The Bergman orthogonal polynomials of Ω satisfy a finite-term recurrence relation, i.e., for every fixed $\ell > 0$, there exists an $N(\ell) > 0$, such that $\langle zp_m, p_\ell \rangle = 0$, $m \geq N(\ell)$.
(5) For any polynomial data there exists a polynomial solution of the Dirichlet problem posed on Ω.

Properties (4) and (5) are essentially equivalent [**152**], and (1) $\Rightarrow$ (2), (2) $\Leftrightarrow$ (3), and (3) $\Rightarrow$ (4). In [**116**] the authors used ratio asymptotics of orthogonal polynomials to show that (2) and equivalently (3) each characterize ellipses. The weaker statement that (1) characterizes ellipsoids was proved in arbitrary dimensions [**137**]. For more about the Khavinson-Shapiro conjecture stated in [**113**], we refer the reader to [**116, 137, 152, 155**] and the references therein.

Thus, if a finite recurrence relation holds for Bergman polynomials for any N, then $N = 3$ and the domain Ω must be an ellipse. In [**116**] this is proved under the assumption that the $\partial\Omega$ is smooth. In the follow up paper [**185**] the result was extended to piece-wise smooth boundaries with corners.

REMARK 18.3. It is well-known [**187**] that *any* family of orthogonal polynomials with respect to an arbitrary measure on the line with uncountable support satisfies a three-term recurrence relation. L. Lempert [**123**] was the first to construct non-trivial examples of C^∞-smooth domains in which Bergman polynomials fail to satisfy any finite recurrence relation. The unexpected connection with the (DP) and (KS)-conjecture revealed [**116, 185**] that any domain that is not an ellipse can serve as such an example.

4. Singularities of the solutions to the Dirichlet problem

As was seen in Chapter 13, our understanding of singularities of solutions to the Cauchy problem of second order equations in two variables is quite complete. In the later Chapters 19-20, we shall also outline a beautiful series of results launched by

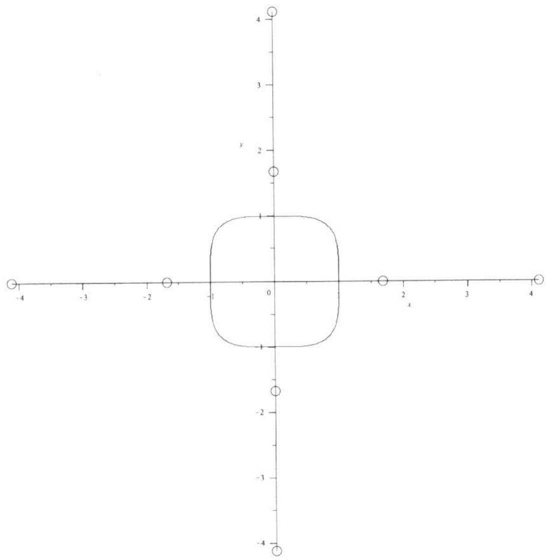

FIGURE 2. A plot of the "TV screen" $\{x^4+y^4=1\}$ along with the first eight singularities (plotted as circles) encountered by analytic continuation of the solution to DP (18.2).

J. Leray discussing the propagation and nature of singularities of the solutions to the Cauchy problem in higher dimensions for the solutions to the Laplace equation. In two dimensions we know that the singularities are closely related to the singularities of the Schwarz function of the initial curve. Thus, it seems natural to assume that the singularities of the solutions to the DP (18.1) outside Ω are somehow associated with the singularities of the Schwarz potential (function) of $\partial\Omega$. However, it turns out that singularities of solutions of the DP are way more complicated than those of the CP. Already in 1992 in his thesis, P. Ebenfelt showed [**37**] that the solution of the following "innocent" DP in $\Omega := \{x^4+y^4-1<0\}$ (the "TV-screen")

$$\begin{cases} \Delta u = 0, \\ u\big|_{\partial\Omega} = x^2+y^2 \end{cases} \tag{18.2}$$

has an infinite discrete set of singularities (of course, symmetric with respect to 90° rotation) sitting on the coordinate axes and running to ∞ (see Figure 2).

To see the difference between analytic continuation of solutions to CP and DP, note that for the former

$$\left.\frac{\partial u}{\partial z}\right|_{\Gamma:=\partial\Omega} = v_z(z,\bar z) = v_z(z,S(z)), \tag{18.3}$$

(where v represents the Cauchy data, i.e., u coincides with v up to first order on Γ), and since $\frac{\partial u}{\partial z}$ is analytic, (18.3) allows u_z to be continued everywhere together with v and $S(z)$, the Schwarz function of $\partial\Omega$. On the other hand, for the DP we have on Γ

$$u(z,\bar z) = v(z,\bar z) \tag{18.4}$$

for $u = f + \bar{g}$ where f and g are analytic in Ω. Hence, (18.4) becomes

$$f(z) + \overline{g(\overline{S(z)})} = v(z, S(z)). \tag{18.5}$$

Now, $v(z, S(z))$ does indeed (for entire v) extend to any domain free of singularities of $S(z)$, but (18.5), even when v is real-valued so that $g = f$, presents a very non-trivial functional equation [**75**] supported by a rather mysterious piece of information that f is analytic in Ω. (18.5) however gives an insight as to how to capture the DP-solution's singularities by considering the DP as part of a Goursat problem in $\mathbb{C}^2$ (or $\mathbb{C}^n$ in general). The latter Goursat problem can be posed as follows (cf. [**175**]).

Given a complex-analytic variety $\hat{\Gamma}$ in $\mathbb{C}^n$, $(\hat{\Gamma} \bigcap \mathbb{R}^n = \Gamma := \partial\Omega)$, find u such that $\sum_{j=1}^n \frac{\partial^2}{\partial z_j^2} u = 0$ near $\hat{\Gamma}$ (and also in $\Omega \subset \mathbb{R}^n$) so that $u\big|_{\hat{\Gamma}} = v$, where v is, say, an entire function of n complex variables. Thus, if $\hat{\Gamma} := \{\phi(z) = 0\}$, where ϕ is, say, an irreducible polynomial, we can, e.g., ponder the following extension of Conjecture 17.2:

Question. For which polynomials ϕ can every entire function v be split (Fischer decomposition) as $v = u + \phi h$, where $\Delta u = 0$ and u, h are entire functions (cf. [**52**], [**175**])?

5. Render's theorem

Trying to establish Conjecture 17.2 H. Render [**155**] has made the following ingenious step. He introduced the *real* version of the Fischer space norm

$$\langle f, g \rangle = \int_{\mathbb{R}^n} f\bar{g}e^{-|x|^2} dx, \tag{18.6}$$

where f and g are polynomials. Originally, the Fischer norm (introduced by E. Fischer [**52**]) requires the integration to be carried over all of $\mathbb{C}^n$ and has the property that multiplication by monomials is adjoint to differentiation with the corresponding multi-index (e.g., multiplication by $(\sum_{j=1}^n x_j^2)$ is adjoint to the differential operator Δ). This property is only partially preserved for the real Fischer norm. More precisely [**155**],

$$\langle \Delta f, g \rangle = \langle f, \Delta g \rangle + 2(deg(f) - deg(g))\langle f, g \rangle \tag{18.7}$$

for homogeneous f, g.

Suppose u solves the DP with data $|x|^2$ on $\partial\Omega \subseteq \{P = 0 : deg(P) = 2k, k > 1\}$. Then $u - |x|^2 = Pq$ for analytic q, and thus $\Delta^k(Pq) = 0$. Using (18.7), this (non-trivially) implies that the real Fischer product $\langle (Pq)_{m+2k}, q_m \rangle$ between all homogeneous parts of degree $m + 2k$ and m of Pq and q, respectively, is zero. By a tour de force argument, Render used this along with an added assumption on the senior term of P (see below) to obtain estimates *from below* for the decay of the norms of homogeneous parts of q. This, in turn yields an if-and-only-if criterion for convergence in the real ball of radius R of the series for the solution $u = \sum_{m=0}^{\infty} u_m$, u_m homogeneous of degree m. Let us state Render's main theorem.

THEOREM 18.2. *Let P be an irreducible polynomial of degree $2k$, $k > 1$. Suppose P is elliptic, i.e. the senior term P_{2k} of P satisfies $P_{2k}(x) \geq c_P|x|^{2k}$, for some constant c_P. Let ϕ be real analytic in $\{|x| < R\}$, and $\Delta^k(P\phi) = 0$ (at least in a neighborhood of the origin). Then, $R \leq C(P, n) < +\infty$, where C is a constant only depending on the polynomial P and the dimension of the ambient space.*

REMARK 18.4. The assumption in the theorem that P is elliptic implies that the set $\{P = 0\}$ is bounded in $\mathbb{R}^n$. The two conditions are not quite equivalent as the example $\{x^2 + y^4 - 1 = 0\}$ shows. The condition that P is *non-hyperbolic* (which is slightly weaker than being elliptic) is equivalent to the boundedness of the set $\{P = 0\}$, see [**155**] for details.

COROLLARY 18.1. *Assume $\partial\Omega$ is contained in the set $\{P = 0\}$, a bounded algebraic set in $\mathbb{R}^n$. Then, if a solution of the DP (18.1) with data $|x|^2$ is entire, Ω must be an ellipsoid.*

PROOF. Suppose not, so $\deg(P) = 2k > 2$, and the following (Fischer decomposition) holds: $|x|^2 = P\phi + u$, $\Delta u = 0$. Hence, $\Delta^k(P\phi) = 0$ and ϕ cannot be analytically continued beyond a *finite* ball of radius $R = C(P) < \infty$, a contradiction. □

REMARK 18.5. We want to stress that although Render's results address both conjectures (KS1) and (KS2) for a large class of domains, both conjectures yet remain open.

Even in two dimensions, both conjectures remain open in general. The conjecture (KS2) even remains open for cubic curves with a bounded (in $\mathbb{R}^2$) component. Settling this case would be a worthy project for a student.

As a step in this direction, we conclude this section by addressing the simpler problem of verifying (KS1) for cubics. This case can be approached using the following result which is a special case of [**137**, Corollary 4.3].

THEOREM 18.3 ([**137**]). *Suppose that ϕ is a polynomial of degree larger than two whose second-highest degree homogeneous part is elliptic. Then the Fischer operator $q \mapsto \Delta(\phi q)$ is not surjective on the space of polynomials.*

Let us apply this result to verify (KS1) in the case of cubics. Assume $\partial\Omega$ is a bounded component of a hypersurface $\{\phi = 0\}$ defined by an irreducible cubic ϕ. Let us show that the second-highest degree term ϕ_2 in the homogeneous expansion $\phi = \phi_3 + \phi_2 + \phi_1 + \phi_0$ may be assumed elliptic. Since Ω is the bounded component of $\phi < 0$, ϕ attains a minimum in Ω. Changing variables with a translation if necessary, we may assume this point (where the minimum occurs) is positioned at the origin. The linear term ϕ_1 is then zero, and the constant term must be negative. If ϕ_2 fails to be elliptic, then it vanishes along some direction $\vec{v}$. In order to ensure that ϕ is eventually positive along this direction (otherwise Ω is unbounded) we must have ϕ_3 positive on both $\vec{v}$ and $-\vec{v}$ which contradicts oddness of ϕ_3. Thus ϕ_2 is elliptic, and we can apply Theorem 18.3 to conclude that the Fischer operator is not surjective, which is equivalent to the existence of polynomial data that does not admit a polynomial solution (see [**158**]), i.e., we have verified that (KS1) holds for cubics.

6. Back to $\mathbb{R}^2$: Annihilating measures and closed lightning bolts

Let us return to the $\mathbb{R}^2$ setting and consider our boundary $\Gamma := \partial\Omega$ of Ω as (part of) an intersection of an analytic Riemann surface $\hat{\Gamma}$ in $\mathbb{C}^2$ with $\mathbb{R}^2$. If say $\partial\Omega$ is a subset of the algebraic curve $\Gamma := \{(x,y) : \phi(x,y) = 0\}$, where ϕ is an irreducible polynomial, then $\hat{\Gamma} = \{(X,Y) \in \mathbb{C}^2 : \phi(X,Y) = 0\}$. Now look at the Dirichlet problem again in the context of the Goursat problem: Given, say, a polynomial data P, find f, g holomorphic functions of one variable near $\hat{\Gamma}$ (a piece of $\hat{\Gamma}$ containing $\partial\Omega \subseteq \hat{\Gamma} \bigcap \mathbb{R}^2$) such that

$$u = f(z) + g(w)\big|_{\hat{\Gamma}} = P(z,w), \tag{18.8}$$

where we have made the linear change of variables $z = X + iY$, $w = X - iY$ (so $\bar{w} = z$ on $\mathbb{R}^2 = \{(X,Y)\colon X, Y \text{ are both real}\}$). Obviously, $\Delta u = 4\frac{\partial^2}{\partial z \partial w} = 0$ and u matches P on $\partial\Omega$. Thus, the DP in $\mathbb{R}^2$ has become an interpolation problem in $\mathbb{C}^2$ of matching a polynomial on an algebraic variety by a sum of holomorphic functions in each variable separately. Suppose that for all polynomials P the solutions u of (18.8) extend as analytic functions to a ball $B_\Omega = \{|z|^2 + |w|^2 < R_\Omega\}$ in $\mathbb{C}^2$. Then, if $\hat{\Gamma} \bigcap B_\Omega$ is path connected, we can interpolate every polynomial $P(z,w)$ on $\hat{\Gamma} \bigcap B_\Omega$ by a holomorphic function of the form $f(z) + g(w)$. Now suppose we can produce a compactly supported measure μ on $\hat{\Gamma} \bigcap B_\Omega$ which annihilates *all* functions of the form $f(z) + g(w)$, f, g holomorphic in B_Ω and at the same time does not annihilate all polynomials $P(z,w)$. Choosing such a $P(z,w)$ (one that fails to be annihilated) as data in (18.8), leads to a solution u that cannot possibly extend to all of the ball B_Ω in $\mathbb{C}^2$; otherwise it would be annihilated by the measure, a contradiction, since u and P coincide at every point of the support. Thus, u must have a singularity in B_Ω in $\mathbb{C}^2$. Then, invoking the theorem of Hayman [**77**] that appeared in Chapter 8 as Corollary 8.2, we would be able to assert that u cannot be extended as a real-analytic function to the *real* disk B_R in $\mathbb{R}^2$ containing Ω with any radius $R \geq \sqrt{2}R_\Omega$.

A simple example of such an annihilating measure, supported at four points (the vertices of a "quadrilateral"), was independently observed by E. Study [**184**], H. Lewy [**131**], and L. Hansen and H. S. Shapiro [**75**]. Namely, assign alternating values ± 1 for the measure supported at the four points $p_0 := (z_1, w_1)$, $q_0 := (z_1, w_2)$, $p_1 := (z_2, w_2)$, and $q_1 := (z_2, w_1)$. Then $\int (f+g)d\mu = f(z_1) + g(w_1) - f(z_1) - g(w_2) + f(z_2) + g(w_2) - f(z_2) - g(w_1) = 0$ for all holomorphic functions f and g of one variable. More generally, the strategy of varying one coordinate and then the other naturally leads to the notion of a *lightning bolt*.

DEFINITION 18.1. A complex *closed lightning bolt* (LB) of length $2(n+1)$ is a finite set of points (vertices) $p_0, q_0, p_1, q_1, \ldots, p_n, q_n, p_{n+1}, q_{n+1}$ such that $p_0 = p_{n+1}$, $q_0 = q_{n+1}$, and each complex line connecting p_j to q_j or q_j to p_{j+1} has either z or w coordinate fixed and they alternate, i.e. if we arrived at p_j with w coordinate fixed then we follow to q_j with z fixed etc.

REMARK 18.6. For "real" domains lightning bolts were introduced by Arnold and Kolmogorov in the 1950s to study Hilbert's 13th problem (see [**117**] and the references therein).

Using the notion of LBs, the following theorem has been proved in [**17**] (see also [**18**]).

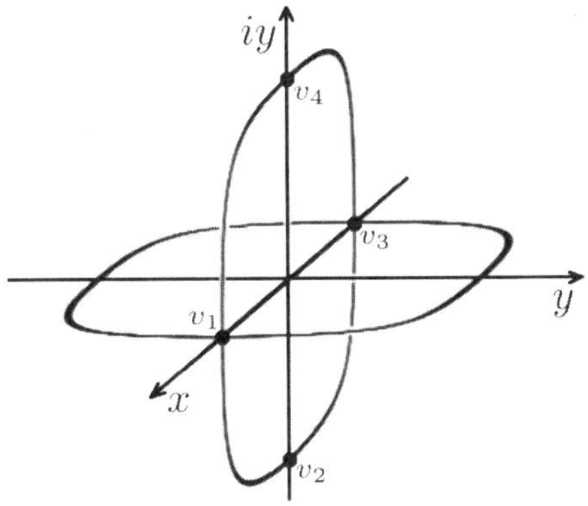

FIGURE 3. A three-dimensional slice of the complexified TV screen $\widehat{\Gamma} := \{(X, Y) \in \mathbb{C}^2 : X^4 + Y^4 = 1\}$ with four vertices v_1, v_2, v_3, v_4 forming a closed lightning bolt.

THEOREM 18.4. *Let Ω be a bounded simply connected domain in $\mathbb{C} \cong \mathbb{R}^2$ such that the Riemann map $\phi : \Omega \to \mathbb{D} = \{|z| < 1\}$ is algebraic. Then all solutions of the DP with polynomial data have only algebraic singularities only at branch points of ϕ with the branching order of the former dividing the branching order of the latter if and only if ϕ^{-1} is a rational function. This in turn is known to be equivalent to Ω being a quadrature domain.*

The complete proof of the theorem is rather technical and uses methods out of the scope of our book, but the idea is rather straightforward.

IDEA OF PROOF: The hypotheses imply that the solution $u = f + \bar{g}$ extends as a single-valued meromorphic function into a $\mathbb{C}^2$-neighborhood of $\hat{\Gamma}$. By another theorem of [**17**], one can find (unless ϕ^{-1} is rational) a continual family of closed LBs on $\hat{\Gamma}$ of bounded length avoiding the poles of u. Hence, the measure with alternating values ± 1 on the vertices of any of these LBs annihilates all solutions $u = f(z) + g(w)$ holomorphic on $\hat{\Gamma}$, but does not, of course, annihilate all polynomials of z, w. Therefore, ϕ^{-1} must be rational, i.e. Ω is a quadrature domain [**175**]. □

Let us revisit Ebenfelt's TV screen example $\{x^4 + y^4 = 1\}$ which turns out to admit a lightning bolt with four vertices on the complexification of $\widehat{\Gamma}$. In (z, w)-coordinates, $\widehat{\Gamma}$ is given by the zero set of $\frac{1}{2}((z+w)^4 + (z-w)^4) - 8 = z^4 + 6z^2w^2 + w^4 - 8$, and we can verify that $\widehat{\Gamma}$ contains the lightning bolt

$$\{(1, 1), (1, -1), (-1, -1), (-1, 1)\}.$$

In terms of the coordinates $X = \frac{z+w}{2}$, $Y = \frac{z-w}{2i}$, the vertices are $v_1 = (1, 0), v_2 = (0, -i), v_3 = (-1, 0)$, and $v_4 = (0, i)$. See Figure 3. The above argument (cf. [**135**]) then leads to the conclusion that the solution to the Dirichlet problem with data $x^2 + y^2$ develops singularities no further from the origin than $2^{3/4}$. Indeed, it can be shown that a real disk of any larger radius has a harmonicity hull that contains not only the closed lightning bolt but also paths along $\widehat{\Gamma}$ connecting the vertices to the real domain (see [**135**]).

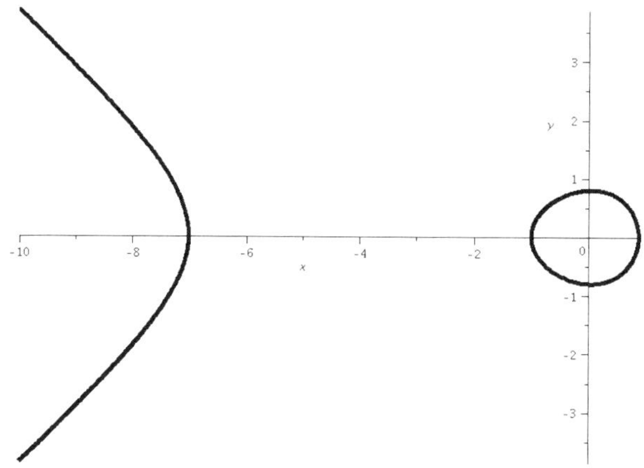

FIGURE 4. A plot of the cubic $8x(x^2-y^2)+57x^2+77y^2-49=0$, showing the bounded component and one unbounded component (there are two other unbounded components further away).

In the same work [**135**], the author constructed other examples of LBs on complexified boundaries of planar domains which do not satisfy the hypothesis of Render's theorem. The LBs validate Conjecture 17.2 and produce an estimate regarding how far into the complement $\mathbb{C} \setminus \overline{\Omega}$ the singularities may develop. For instance, the complexification of the cubic, $8x(x^2-y^2)+57x^2+77y^2-49=0$ has a lightning bolt with six vertices in the plane where z and w are real, i.e., x is real and y is imaginary (see Figure 4 for a plot of the cubic in the plane where x and y are real and see Figure 5 for the slice containing the lightning bolt).

Repeating the method outlined in the above arguments, we conclude that if the solution with appropriate cubic data is analytically continued in the direction of the closest unbounded component of the curve defining $\partial\Omega$, it must develop a singularity before it can be forced to match the data on that component.

This confirms an intuition that matching data on additional components represents an overdetermined problem, thus presenting an obstruction that forces singularities to appear as soon as the solution is continued to a sufficiently large disk. More precisely, we are led to conjecture the following.

CONJECTURE 18.1. *Suppose Ω is a bounded domain containing the origin with the boundary $\partial\Omega$ being a component of a smooth, irreducible real algebraic hypersurface $\Gamma \subset \mathbb{R}^n$ that has at least one additional connected component. If the harmonicity hull $\widehat{B}_R \subset \mathbb{C}^n$ of the ball $B_R := \{|x| < R\}$ contains a path along the complexification $\widehat{\Gamma} \subset \mathbb{C}^n$ joining $\partial\Omega$ to another component of Γ then there exist polynomial data for which the solution to the Dirichlet problem (posed in Ω) develops singularities in B_R.*

We note that any irreducible algebraic variety $\widehat{\Gamma}$ is connected (and path connected) in $\mathbb{C}^n$, even if the real slice Γ consists of several components. Hence, the condition on $\widehat{B}_R$ holds for all R sufficiently large.

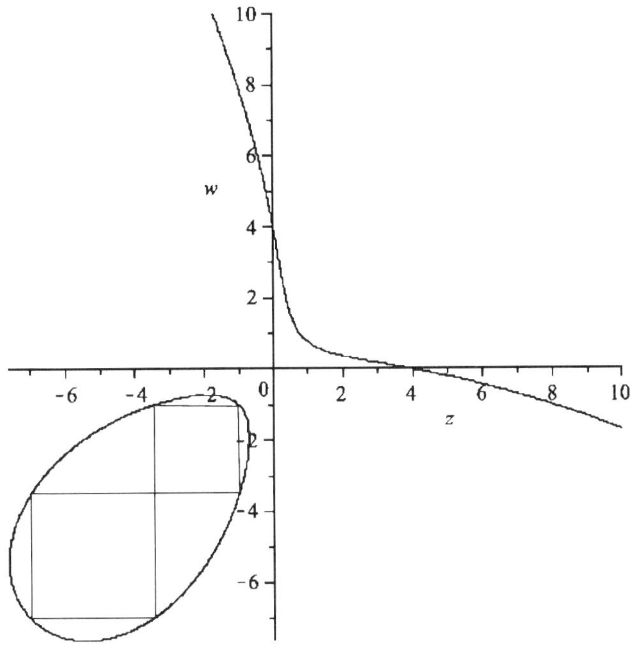

FIGURE 5. A lightning bolt with six vertices on the cubic $2(z+w)(z^2+w^2)+67zw-5(z^2+w^2)=49$ in the plane where z and w real, i.e. x is real and y is imaginary.

Notes

In two dimensions, one of the main results in [**17**] yields that disks are the only domains for which all solutions of the DP with rational (in x, y) data v are rational. On the other hand, algebraic data may lead to a transcendental solution even in disks (see [**43**], also cf. [**44**]). In three dimensions and higher, rational data on the sphere (e.g., $v = \frac{1}{x_1 - a}$, $|a| > 1$) yield transcendental solutions of (18.1), although no estimate on the precise location of singularities of that transcendental solution exists as of now (cf. [**43**]).

For further discussion of singularities of solutions to the Dirichlet problem, as well as the Khavinson-Shapiro conjectures, we refer the reader to the surveys [**105**], [**156**]. The more general setting of studying the continuation properties of solutions of Dirichlet problem for higher order operators, e.g., the bilaplacian, can be traced back to [**78**]; also cf. [**155**] for further relevant discussion.

It is still not clear on an intuitive level why ellipsoids play such a distinguished role in providing "excellent" solutions to DP with "excellent" data. A very similar question, important for applications, (which actually inspired the program launched in [**113**] on singularities of the solutions to the DP) goes back to Raleigh and concerns singularities of solutions of the Helmholtz equation ($[\Delta - \lambda^2]u = 0$, $\lambda \in \mathbb{R}$) instead. (The minus sign will guarantee that the maximum principle holds and, consequently, ensures uniqueness of solutions of the DP.) To the best of our knowledge, this topic remains virtually unexplored.

An affirmative answer to the Conjecture 18.1 would confirm both KS1 and KS2 for a large class of domains, addressing essentially all of the smooth irreducible algebraic boundaries left open by Render's theorem. Indeed, if Render's ellipticity condition is not satisfied then, except for degenerate cases, the real algebraic set is unbounded and hence contains more than one component (a bounded component to serve as $\partial\Omega$ as well as at least one unbounded component).

CHAPTER 19

An Introduction to J. Leray's Principle on Propagation of Singularities through $\mathbb{C}^n$

1. Introductory remarks on propagation of singularities

Let us return to the discussion of the questions posed in the introduction in Examples 1.3 and 1.2 from Sec. 1 and Question E from Sec. 5 of Chapter 1.

The main purpose of this and the following chapter is to open the door for graduate students and young researchers to a circle of crucial and beautiful ideas of J. Leray concerning local propagation of singularities of solutions of the holomorphic Cauchy problem, and further development of these ideas by G. Johnsson in order to obtain information on *global* behavior of solutions of some Cauchy problems with entire data on quadric surfaces in $\mathbb{C}^n$. We restrict ourselves to the case of the Laplace operator and in some cases to certain simple first-order operators. In order to keep the exposition transparent, we replace most proofs by detailed treatment of particular examples. We stress again that our only (modest) goal here is to present an outline of the most important ideas and point out the corresponding references.

First, let us recall the basic setup for the general Cauchy problem and the most fundamental notions associated with it (cf. [**61**], [**72**], [**86**], [**150**], etc.). We have a complex non-singular holomorphic hypersurface Γ in $\mathbb{C}^n$ of complex dimension $n-1$, i.e., $\Gamma := \{z \in \mathbb{C}^n : \varphi(z) = 0\}$, where φ is a function holomorphic in a certain neighborhood V of $z^0 : \varphi(z^0) = 0$ and the "complex gradient" $\nabla\varphi := (\partial_1\varphi_1, \ldots, \partial_n\varphi) \neq (0, 0, \ldots, 0)$ in V $(\partial_i = \partial/\partial z_i)$.

Let $\mathcal{L} := \sum_{|\alpha|\leq m} a_\alpha(z)\partial^\alpha$ be the most general linear differential operator with coefficients a_α holomorphic in V, and consider the "Cauchy problem" to find a function u holomorphic on a neighborhood of z^0 such that

$$\left\{\begin{aligned} \mathcal{L}u(z) &= \sum_{|\alpha|\leq m} a_\alpha(z)\partial^\alpha u(z) = g(z) \text{ near } z^0, \\ u &\equiv f \text{ on } \Gamma, \text{ i.e., } \partial^\alpha(u-f) = 0 \text{ on } \Gamma \text{ for all } \alpha : |\alpha| \leq m-1, \end{aligned}\right. \tag{19.1}$$

where g and "data" f are also holomorphic in, say, V.

Recall the notion of a characteristic point.

DEFINITION 19.1. Γ is said to be non-characteristic with respect to $\mathcal{L}$ at z^0 if

$$\sum_{|\alpha|=m} a_\alpha (\nabla\varphi)^\alpha (z^0) := \sum_{|\alpha|=m} a_\alpha(z) \left(\frac{\partial\varphi}{\partial z_1}\right)^{\alpha_1} \cdots \left(\frac{\partial\varphi}{\partial z_n}\right)^{\alpha_n}\Bigg|_{z^0} \neq 0. \tag{19.2}$$

DEFINITION 19.2. A point z^0 on Γ is called a characteristic point with respect to $\mathcal{L}$ if the left-hand side of (19.2) vanishes at z^0. The totality of such points on Γ is denoted Γ_{ch}.

DEFINITION 19.3. A variety Γ is called a characteristic variety for $\mathcal{L}$ if every point on Γ is characteristic.

The Cauchy-Kovalevskaya Theorem (cf. Ch. 2) states that if $z^0 \notin \Gamma_{ch}$, a solution u of (19.1) exists in some neighborhood of z^0; moreover, u is unique.

EXAMPLE 19.1.

(i) $\mathcal{L} = z_1 \frac{\partial^2}{\partial z_2^2} - \frac{\partial^2}{\partial z_1^2}$; $(0,0)$ is a characteristic point of $\{z_2 = 0\}$.

(ii) The plane $\{z_2 = 0\}$ is an everywhere characteristic variety for the operator $\mathcal{L} = z_2^2 \frac{\partial^2}{\partial z_2^2} - \frac{\partial^2}{\partial z_1^2}$.

(iii) That characteristic points mean trouble is immediately clear if, e.g., we consider for $\mathcal{L}$ as in (i) the Cauchy problem (19.1) with Cauchy data $f = 0$ on $\Gamma = \{z_2 = 0\}$. Then along $\{z_2 = 0\}$ we have that the solution u, and the partial derivatives $\partial u/\partial z_1$, $\partial^2 u/\partial z_1^2$ vanish, and hence if $g(0,0) \neq 0$ then there is no solution.

(iv) An example of non-uniqueness near characteristic points is given by

$$\mathcal{L} = \sum_1^n \partial^2/\partial z_j^2 \text{ and } \Gamma = \left\{ z \in \mathbb{C}^n : \sum_1^n a_j z_j = 0, \sum a_j^2 = 0 \right\},$$

an everywhere characteristic (with respect to $\mathcal{L}$) hyperplane. Then $u = \left(\sum_1^n a_j z_j \right)^k$, for arbitrary $k \geq 2$, is a non-trivial solution of the Cauchy problem (19.1) with zero data.

The latter example is a simple partial case of the following result due to Goursat (cf. [**85**, Thm. 5.1.3] and also [**71**, p. 105-107]). Before stating the theorem, we need one more technical definition. Denote by $P(z,\xi) = \sum_{|\alpha|=m} a_\alpha(z)\xi^\alpha$ the principal symbol of the operator $\mathcal{L}$, whose coefficients $a_\alpha(z)$ are holomorphic in a neighborhood of z^0. Let $\Gamma : \{z \in \mathbb{C}^n : \varphi(z) = 0\}$ be a piece of a non-singular holomorphic hypersurface which is characteristic at z^0 for $\mathcal{L}$, i.e., satisfies $P\left(z^0, \nabla\varphi\left(z^0\right)\right) = 0$.

DEFINITION 19.4. z^0 is simply characteristic (or, Γ is simply characteristic at z^0) if, denoting $P_j = \partial P/\partial \xi_j$, we have $P_j\left(z^0; \nabla\varphi\left(z^0\right)\right) \neq 0$ for at least one $j \in \{1, 2, \ldots, n\}$.

For example, for $\mathcal{L} = \sum_1^n \partial^2/\partial z_j^2$, Γ is characteristic at z^0 if and only if

$$\sum_{j=1}^n \left(\partial\varphi/\partial z_j\right)^2 = 0$$

at z^0; z^0 is simply characteristic if $\partial\varphi/\partial z_j \neq 0$ at z^0 for some j, i.e., in this case "simply characteristic" just means "characteristic and non-singular." (In case all $P_j\left(z^0, \nabla\varphi\left(z^0\right)\right)$ vanish but $P_{jk}\left(z^0; \nabla\varphi\left(z^0\right)\right) \neq 0$ for some j, k we say $\mathcal{L}$ is doubly characteristic at z^0, and so on.)

THEOREM 19.1 (Goursat). *Let $\Gamma = \{z \in \mathbb{C}^n : \varphi(z) = 0\}$ be a (piece of a) non-singular holomorphic hypersurface. Suppose, moreover, that Γ is simply characteristic for the linear partial differential operator*

$$\mathcal{L} = \sum_{|\alpha| \leq m} a_\alpha(z)\partial^\alpha$$

with coefficients holomorphic in a neighborhood of Γ. *Then, for every integer* $k > 0$ *there exists a solution* u *of* $\mathcal{L}u = 0$, *holomorphic near* z^0, *not vanishing identically, and such that* $\partial^\alpha u = 0$ *on* Γ, *for all* $\alpha : |\alpha| \leq k$.

For example, taking $\mathcal{L} = \sum_1^{n-1} \partial^2/\partial z_j^2 - \partial^2/\partial z_n^2$ and the (complex) "cone" $\Gamma = \left\{ z \in \mathbb{C}^n : \sum\limits_1^{n-1} z_j^2 - z_n^2 = 0 \right\}$, we find that $\Gamma \setminus \{0, 0, \ldots, 0\}$ is everywhere simply characteristic and therefore, by Goursat's theorem we find a non-trivial solution of $\mathcal{L}u = 0$ near an arbitrary piece of this cone not containing the origin, which vanishes to any desired order on the cone. (Intersecting this picture with $\mathbb{R}^n$, we get a corresponding result for the real wave equation and a "light cone.")

Note that we must remove the origin before applying Goursat's theorem in the above example, since Γ is singular there. To illustrate that this assumption is needed, let us suppose that a harmonic function u extends to a neighborhood in $\mathbb{C}^n$ of the isotropic cone $\{z \in \mathbb{C}^n : \sum_1^n z_j^2 = 0\}$ (which is everywhere-characteristic with respect to the Laplacian), and suppose u vanishes on the isotorpic cone but does not vanish identically. By Corollary 5.1 u extends to all of $\mathbb{C}^n$. We can then apply Theorem 1 in [**175**] in order to obtain that u is divisible by $P(z) := \sum_1^n z_j^2$, and this implies that each homogeneous term in the series expansion of u is divisible by P. Since $P(z)$ is nonnegative for $z \in \mathbb{R}^n$, this violates Theorem 17.2 (the Brelot-Choquet Theorem). For the proof of Goursat's theorem and its generalizations, we refer the reader to, e.g., [**85**], [**71**].

Although the Cauchy-Kovalevskaya theorem says that a solution of (19.1) exists on a neighborhood of $\Gamma \setminus \Gamma_{ch}$, this does not tell the whole story. Typically, a solution u exists outside a certain characteristic variety which is tangent to Γ along Γ_{ch}. This is called "Leray's principle" and will be discussed in Section 2.

Here, we want to address a very natural question arising in connection with "Leray's principle"; which varieties can be carriers of singularities? The answer is given by a beautiful theorem of Delassus and Le Roux (cf. [**72**, Ch. 3]). Note that we have already seen its simplified version in Sec. 4 of Ch. 4, though coached in slightly different context to avoid use of Hartogs' theorem there.

THEOREM 19.2 (Delassus, Le Roux). *Let* Γ *be a non-singular holomorphic hypersurface, and suppose for some* $z^0 \in \Gamma$ *that* u *solves* $\mathcal{L}u = g$ *in* $B \setminus \Gamma$, *where* B *is a ball in* $\mathbb{C}^n$ *centered at* z^0. *Then* u *is holomorphically continuable across* Γ *unless* Γ *is an everywhere characteristic variety for* $\mathcal{L}$. *In consequence, if* u *is not continuable across some point of* Γ, *then* u *is singular on the whole portion of* Γ *in* B, *and* Γ *is everywhere characteristic for* $\mathcal{L}$.

For a classical "calculus" proof of a slightly weaker version of the theorem, we refer the reader to Hadamard's book [**72**]. Here, following Johnsson [**94**] we give a very short proof relying on Zerner's theorem (cf. Ch. 4) and an old theorem of Hartogs [**14**, Satz 19].

Keeping the same notation as in Theorem 19.2 we have

THEOREM 19.3 (Hartogs). *Let* u *be a holomorphic function in* $B \setminus \Gamma$ *(not necessarily single-valued, i.e., it is continuable indefinitely along all paths in* B *not meeting* Γ*). Then, if* u *is continuable across some point* $z' \in \Gamma$, *it is continuable across all points on* $\Gamma \cap B$.

Now we can at once prove Theorem 19.2.

PROOF OF THEOREM 19.2. Let $\Gamma = \{z \in B : \varphi(z) = 0\}$, φ being holomorphic and non-singular on B. Consider the real C^1-hypersurface

$$S = \{z \in \mathbb{C}^n : v(z) := Re\, \varphi(z) = 0\}.$$

Then S contains Γ and the Cauchy-Riemann equations yield

$$\frac{\partial v}{\partial z_j} = \left(\frac{1}{2}\right)\left(\frac{\partial v}{\partial x_j} - i\,\frac{\partial v}{\partial y_j}\right) = \left(\frac{1}{2}\right)\left(\frac{\partial \varphi}{\partial z_j} + \frac{\partial \varphi}{\partial \bar{z}_j}\right) = \left(\frac{1}{2}\right)\frac{\partial \varphi}{\partial z_j},$$

i.e., S is Zerner-characteristic at some point of $\Gamma \subset S$ if and only if Γ is characteristic there. Now if u is singular at a point of $B \cap \Gamma$, it is singular on all of $B \cap \Gamma$ by Hartogs' theorem, and $\Gamma \subset S$ must be everywhere characteristic (with respect to $\mathcal{L}$) by Zerner's theorem (Thm. 4.1), since u is holomorphic in $B \setminus \Gamma \supset B \setminus S$. $\square$

EXAMPLE 19.2. $u = \left(\sum_1^n z_j^2\right)^{(2-n)/2}$ satisfies the Laplace equation

$$\sum_1^n \partial^2 u/\partial z_j^2 = 0$$

in $\mathbb{C}^n \setminus \Gamma_0$, where $\Gamma_0 = \left\{z \in \mathbb{C}^n : \sum_1^n z_j^2 = 0\right\}$ is an isotropic cone and u is singular everywhere on Γ_0, which is everywhere characteristic in conformity with Theorem 19.2.

Thus, generically the "vehicle for propagation of singularities" of solutions to holomorphic partial differential equations is everywhere characteristic varieties. Theorem 19.2 has a converse. This is stated in Hadamard's book [**72**] and ostensibly proved, although we haven't been able to verify the proof.

A complete proof of a partial converse was given fairly recently by Tsuno [**193**] in the following version.

THEOREM 19.4. *Let* $\mathcal{L} = \sum_{|\alpha| \le m} a_\alpha(z)\partial^\alpha$ *be a linear operator (where, as usual,* a_α *are holomorphic near some point* z^0*) and suppose* Γ *is a non-singular everywhere characteristic variety passing through* z^0 *which is, moreover, simply characteristic. Then there is an open ball* B *centered at* z^0*, and a function* u *holomorphic on* $B \setminus \Gamma$ *and satisfying* $\mathcal{L}u = 0$ *there, such that* u *is not analytically continuable across any point of* Γ*; i.e.,* Γ *is precisely the set of singularities of* u*.*

With those remarks, we close this introductory section and shall have a glimpse of Leray's theory.

2. Local propagation of singularities in $\mathbb{C}^n$: Leray's principle

Let $\mathcal{L} = \sum_{|\alpha| \le m} a_\alpha(z)\partial^\alpha$ be a differential operator, $P_m(z, \xi) = \sum_{|\alpha| = m} a_\alpha(z)\xi^\alpha$ its principal symbol.

DEFINITION 19.5. The integral curves $z = z(t, \xi)$, ξ being a tangent vector at point $z = z(t)$, $t \in \mathbb{C}$ of the Hamilton-Jacobi system of equations ($z = (z_1, \ldots, z_n) \in$

$\mathbb{C}^n$, $\xi = (\xi_1, \ldots, \xi_n) \in \mathbb{C}^n$)

$$\text{(19.3)} \qquad \begin{cases} \dfrac{dz_j}{dt} = \dfrac{\partial P_m}{\partial \xi_j}, & j = 1, \ldots, n, \\[2ex] \dfrac{d\xi_j}{dt} = -\dfrac{\partial P_m}{\partial z_j}, & j = 1, \ldots, n, \end{cases}$$

$$z(0) = z^0, \quad \xi(0) = \xi^0$$

are called bicharacteristics (of $\mathcal{L}$), emanating from the point z^0 and "tangent" to the plane "orthogonal" to the vector ξ^0 at z^0, (when we say that vectors $a, b \in \mathbb{C}^n$ are "orthogonal," we mean $a \cdot b := \sum_{i=1}^{n} a_i b_i = 0$.)

For more details, we refer the reader to detailed treatises by Garabedian [**61**], Hadamard [**72**], Hörmander [**86**], John [**93**], etc. Although the formalism is the same, we stress that here t and z are complex.

EXAMPLE 19.3. $\mathcal{L} = \sum_1^n \partial^2/\partial z_j^2$, $P_m = \sum_1^n \xi_i^2$ and equations (19.3) transform into

$$\begin{cases} \dfrac{dz}{dt} = 2\xi, \\[2ex] \dfrac{d\xi}{dt} = 0. \end{cases}$$

Then, the solutions $z = z^0 + 2t\xi^0$ are in fact complex lines.

The following basic notion is due to Leray [**126**], who does not, however, use the same terminology.

DEFINITION 19.6. Let $\Gamma = \{z : \varphi(z) = 0\}$ be a holomorphic hypersurface non-singular at z^0. A bicharacteristic (Def. 19.5) is called a *principal bicharacteristic* with respect to Γ if ξ^0 is taken to be $(\nabla_z \varphi)(z^0)$.

EXAMPLE 19.4 (Example 19.3 continued). Observe that if Γ is characteristic with respect to $\mathcal{L}$ at z^0 the principal bicharacteristic is "tangent" to Γ at z^0. For $n = 3$, this already determines it as the unique complex line along which the isotropic cone $\Gamma_{z^0} := \left\{ z \in \mathbb{C}^3 : \sum_1^3 \left(z_i - z_i^0\right)^2 = 0 \right\}$ "touches" Γ, while for $n \geq 4$ further specifications are needed.

The generic picture of propagation of singularities in $\mathbb{C}^n$ is, roughly speaking, as follows.

LERAY'S PRINCIPLE 19.1. [cf. [**126**]] Singularities are "born" at the points of Γ_{ch} and propagate to $\mathbb{C}^n$ near Γ along principal bicharacteristics emanating from non-exceptional points of Γ_{ch}, those bicharacteristics forming an everywhere characteristic variety Σ tangent to Γ along Γ_{ch}.

Before explaining what non-exceptional points are and how Leray's method leads to the above principle, let us take up an instructive example. For the sake of clarity we even place it in $\mathbb{R}^2$ rather than $\mathbb{C}^2$.

EXAMPLE 19.5. Let $\Gamma = \left\{x \in \mathbb{R}^2 : x_2 = f(x_1)\right\}$ be a graph of a real-analytic function, and consider the Cauchy problem

$$\left\{\begin{aligned} \frac{\partial^2 u}{\partial x_1 \partial x_2} &= 1 \text{ near } \Gamma, \\ u &\equiv 0 \text{ on } \Gamma, \end{aligned}\right. \tag{19.4}$$

where we recall that the notation $u \equiv 0$ on Γ means that both u and ∇u vanish on Γ. Assume $f(0) = 0 = f'(0)$, so the origin $(0,0)$ is in fact a characteristic point of Γ. There are two bicharacteristics passing through $(0,0)$, the tangent one $\{x_2 = 0\}$ and the line $\{x_1 = 0\}$ transversal to Γ. The solution of (19.4) has the form

$$u = x_1 x_2 + a(x_1) + b(x_2), \tag{19.5}$$

where we must choose a, b so that

$$\left\{\begin{aligned} x_1 x_2 + a(x_1) + b(x_2) &= 0 \text{ on } \Gamma, \\ x_2 + a'(x_1) &= 0 \text{ on } \Gamma, \\ x_1 + b'(x_2) &= 0 \text{ on } \Gamma, \end{aligned}\right.$$

i.e., $a'(x_1) = -f(x_1)$ and $b'(x_2) = f^{-1}(x_2)$, where f^{-1} is the inverse of f and hence is singular at $x_2 = 0$. Thus, in (19.5) the singular term is $b(x_2)$ and the singularity propagates along the (principal) tangent bicharacteristic only. (Since for the first-order operators in two variables bicharacteristics and characteristics turn out to be the same thing, the line $\{x_2 = 0\}$ is, in fact, the tangent characteristic surface as well.)

The reader may have noticed that the solution of (19.4) is nothing else than the modified Schwarz potential of Γ, and it develops a singularity as it does at *all* characteristic points in accordance with Shapiro's theorem (Prop. 13.3, also, see [**94**]).

Now let us describe briefly Leray's general uniformization procedure leading to Leray's Principle 19.1. Here we are following [**94**], which is a somewhat simplified version of [**61**]. (For a more general setup dealing with systems, we refer the reader to [**62**].) Thus, we seek the holomorphic function u solving

$$\left\{\begin{aligned} \mathcal{L}u &= g, \quad \mathcal{L} = \sum_{|\alpha|\le m} a_\alpha(z)\partial^\alpha, \\ u &\equiv f \text{ on } \Gamma \setminus \Gamma_{ch}, \end{aligned}\right. \tag{19.6}$$

where $\Gamma = \{z \in \mathbb{C}^n : \varphi(z) = 0\}$ is a holomorphic hypersurface and a_α, f, g are holomorphic in a neighborhood of Γ.

The first step transforms (19.6) to a related problem in $\mathbb{C}^{n+1}$ by introducing a new (complex) variable λ. Set

$$S^\lambda := \left\{(z,\lambda) \in \mathbb{C}^{n+1} : \varphi(z) = \lambda\right\},$$

for small $|\lambda|$, and consider a new Cauchy problem in $\mathbb{C}^{n+1}$;

$$\left\{\begin{aligned} &\sum_{|\alpha|\le m} a_\alpha(z)(\partial/\partial z)^\alpha u^*(z,\lambda) = g(z) \text{ near } S^\lambda \setminus S^\lambda_{ch}, \\ &\partial^\beta_{z,\lambda}\left(u^* - f(z)\right) = 0 \text{ on } S^\lambda \setminus S^\lambda_{ch} \text{ for all multi-indices } \beta : |\beta| \le m-1. \end{aligned}\right. \tag{19.7}$$

(Here ∂_z, $\partial_{z,\lambda}$ denote differentiation with respect to the indicated variables.) Note that

$$u(z) = u^*(z,0). \tag{19.8}$$

The second step consists of introducing a new parameter t in place of λ by setting $\lambda = \lambda(z,t)$, where λ is defined as the solution of the following first-order initial-value problem

$$\left\{ \begin{array}{l} \lambda_t = -P_m\left(z; \nabla_z \lambda\right), \\ \lambda(z,0) = \varphi(z), \end{array} \right. \tag{19.9}$$

$(P_m(z;\xi) = \sum\limits_{|\alpha|=m} a_\alpha(z)\xi^\alpha$ as before). Solving (19.9) by means of the classical Hamilton-Jacobi theory (cf. [**93**], also see [**62**] and [**61**]) we find that

$$\lambda(z,t) = \lambda(w(z,t),t), \tag{19.10}$$

where

$$\lambda(w,t) = \varphi(w) + t(m-1)P_m\left(w; \nabla_w \varphi(w)\right) \tag{19.11}$$

and $w = w(z,t)$ is defined implicitly by the relation

$$z = z(w,t), \tag{19.12}$$

which represents the solution of the following Hamilton-Jacobi equations

$$\left\{ \begin{array}{ll} \dfrac{dz}{dt} & = \dfrac{\partial P_m}{\partial \xi}, \\ \dfrac{d\xi}{dt} & = -\dfrac{\partial P_m}{\partial z}, \\ z(0) & = w, \quad \xi(0) = \nabla_w \varphi, \end{array} \right. \tag{19.13}$$

i.e., the curves $z = z(t)$ are principal bicharacteristics emanating from w. Observe now that in view of (19.9) under the transformation $\lambda = \lambda(z,t)$, the initial surface S^λ in (19.7) becomes a hyperplane

$$Z_t = \left\{(z,t) \in \mathbb{C}^{n+1} : t = 0\right\}.$$

Set $U(z,t) = u^*(z, \lambda(z,t))$ and $V(z,t) = \left(\partial u^*/\partial\lambda\right)(z, \lambda(z,t))$. It is left as a routine but instructive exercise for the reader to check directly using (19.7 - 19.13) that the so-defined function V is a solution of the non-characteristic Cauchy problem

$$\left\{ \begin{array}{rl} \sum\limits_{|\beta|\le m} b_\beta(z,t)\partial^\beta_{z,t}V(z,t) & = 0 \text{ near } Z_t, \\ V & \equiv f_1 \text{ on } Z_t, \end{array} \right. \tag{19.14}$$

where f_1 is holomorphic near Z_t (once again, we recall that $\equiv$ means that the functions and all their derivatives up to order $(m-1)$ coincide on Z_t).

It is now clear that a solution $V(z,t)$ of (19.14) yields a solution u of (19.6). Indeed, first $U(z,t)$ can be determined by

$$\begin{aligned} U(z,t) &= U(z,0) + \int_0^t \frac{\partial U}{\partial s}(z,s)\,ds = U(z,0) + \int_0^t V(z,s)\,\lambda_s(z,s)\,ds \\ &= f(z) + \int_0^t V(z,s)\,\lambda_s(z,s)\,ds, \end{aligned} \tag{19.15}$$

since $U(z,0) \stackrel{\text{def}}{=} u^*(z,\lambda(z,0)) = u^*(z,\lambda)|_{S^\lambda} = f(z)$ by (19.7). Secondly, since $u(z) = u^*(z,0)$ by (19.8) and $u^*(z,\lambda(z,t)) = U(z,t)$ by definition, we have

$$u(z) = U(z,t(z)), \tag{19.16}$$

where $t(z)$ is defined as an implicit function from $\lambda(z,t) = 0$. From (19.14)–(19.16) and the Cauchy-Kovalevskaya theorem it is seen at once that sufficiently close to Γ

the solution u of (19.6) becomes singular only at those points where $t(z)$ does not extend holomorphically, i.e., where the equation $\lambda(z,t)=0$ cannot be solved with respect to t. Such points can be calculated by eliminating t from the equations

$$\left\{\begin{aligned} \lambda(z,t) &= 0, \\ \frac{\partial}{\partial t}\lambda(z,t) &= 0. \end{aligned}\right. \tag{19.17}$$

Equations (19.17) define a so-called envelope Σ of the 1-parameter family of surfaces $\Gamma_t = \{z : \lambda(z,t)=0\}$, i.e., a surface Σ tangent to all Γ_t.

By a standard application of the Hamilton-Jacobi theory (cf. [**62**], [**126**]) one can verify that the surface Σ is

(i) everywhere characteristic (with respect to $\mathcal{L}$);
(ii) tangent to Γ along Γ_{ch};
(iii) contained in a union of all principal bicharacteristic β_z, $z \in \Gamma_{ch}$.

Thus, near the "ordinary" characteristic points, i.e., points for which the argument (19.14–13) holds, all singularities of the solution u of (19.7) lie on the characteristic tangent Σ. Now let us examine the possible exceptional points where the above argument does not work. One obstacle is the existence of the λ-function, the uniformizing parameter defined as a solution of (19.9). As is easily seen from the standard sufficient condition for the existence of solutions of first-order equations (cf. e.g., [**93**, Ch. I]), assuming Γ to be simply characteristic at z^0 guarantees existence of the solution $\lambda(z,t)$ of (19.9) near z^0. The second more serious obstacle can occur if the equation $\lambda(z,t)=0$ is not solvable with respect to z in a whole neighborhood of the origin $t=0$ in the t-plane, i.e., for some $z^0 \in \Gamma_{ch}$, $\lambda\left(z^0,t\right)=0$ in a neighborhood of $t=0$ and it is not hard to show ([**62**]) that in that case the principal bicharacteristic $\beta_{z^0} \subset \Gamma_{ch}$ in a whole neighborhood of z^0 and, therefore, the above procedure does not tell us anything at all about the propagation of singularities of u off Γ. We shall call such a point $z^0 \in \Gamma_{ch}$ an exceptional point.

Thus we can now reformulate the heuristic Leray's Principle 19.1 into a precise theorem also due to Leray.

THEOREM 19.5 (cf. [**126**], [**62**]). *Consider the Cauchy problem* (19.6)*, where the initial surface* Γ *is assumed to be non-singular and simply characteristic at* $z^0 \in \Gamma$*. Assume, moreover, that* z^0 *is not an exceptional point of* Γ*. Then* (19.6) *has a* (*unique*) *holomorphic solution* (*perhaps multivalued*) *in a neighborhood* Ω *of* z^0*, except for points on the characteristic tangent* Σ *in* Ω.

REMARK 19.1. Hamada [**73**] has extended the above result by replacing the requirement for Γ to be simply characteristic at z^0 by multiple (finite) characteristicity at z^0. He showed that the conclusion of Theorem 19.5 still holds even then; however, the singularities on Σ may be essential. (In view of Weierstrass' Preparation theorem, the condition that z^0 is not an exceptional point implies that the equation $\lambda(z,t)=0$ produces an "algebroid function" $z \mapsto t(z)$ and, therefore, the singularity of $u(z)$ near z^0 is a bounded singularity.)

We finish this section with three examples (cf. [**94**]). In the first example we exhibit the uniformization procedure (19.6)–(19.14) for the Laplace operator. Then in the second we apply it to a particular Cauchy problem. Finally, in the third example we show Leray's theory failing in an exceptional case.

EXAMPLE 19.6. Suppose we are looking for a solution of the Cauchy problem

$$\begin{cases} \Delta u &= g \quad \left(\Delta = \sum_1^n \partial^2/\partial z_j^2\right), \\ u &\equiv f \text{ on } \Gamma \setminus \Gamma_{ch} \end{cases} \tag{19.18}$$

near a holomorphic surface $\Gamma := \{z \in \mathbb{C}^n : \varphi(z) = 0\}$. Then (19.7) becomes

$$\begin{cases} \Delta u^*(z,\lambda) &= g(z), \\ u^*(z,\lambda) &= f(z) \text{ on } S^\lambda := \{(z,\lambda) \in \mathbb{C}^{n+1} : \varphi(z) = \lambda\}. \end{cases} \tag{19.19}$$

Set

$$\begin{aligned} &\text{(i)} \quad U(z,t) = u^*(z, \lambda(z,t)), \\ &\text{(ii)} \quad V(z,t) = u^*_\lambda(z, \lambda(z,t)), \\ &\text{(iii)} \quad W(z,t) = u^*_{\lambda\lambda}(z, \lambda(z,t)), \end{aligned} \tag{19.20}$$

where λ solves (cf. (19.9))

$$\begin{cases} \lambda_t = -\sum_{j=1}^n \lambda_j^2, \quad (\lambda_j \text{ denotes } \lambda_{z_j}, \text{ etc.}), \\ \lambda(z,0) = \varphi(z). \end{cases} \tag{19.21}$$

Differentiating (19.20i) we find

$$U_j = u_j^* + u_\lambda^* \lambda_j, \text{ i.e., } u_j^* = U_j - V\lambda_j. \tag{19.22}$$

Also,

$$U_{jk} = u_{jk}^* + u_{j\lambda}^* \lambda_k + (u_{\lambda\lambda}^* \lambda_k + u_{k\lambda}^*)\, \lambda_j + u_\lambda^* \lambda_{jk}. \tag{19.23}$$

Differentiating (19.20ii) with respect to z_j we obtain

$$V_j = u_{\lambda j}^* + u_{\lambda\lambda}^* \lambda_j = u_{\lambda j}^* + W\lambda_j.$$

So,

$$u_{\lambda j}^* = V_j - W\lambda_j$$

and

$$u_{jk}^* = U_{jk} - V_j\lambda_k - V_k\lambda_j - V\lambda_{jk} + W\lambda_j\lambda_k. \tag{19.24}$$

Hence, the equation in (19.19) becomes

$$\Delta U - 2\sum_{j=1}^n V_j\lambda_j - V\Delta\lambda + W\sum_{j=1}^n \lambda_j^2 - g = 0. \tag{19.25}$$

Now, in view of (19.21) and (19.20ii), we have

$$W\sum_{j=1}^n \lambda_j^2 = -W\lambda_t = -V_t.$$

Hence from (19.25) and (19.20i) we obtain the system

$$\begin{cases} V_t = \Delta U - 2\sum_{j=1}^n V_j\lambda_j - V\Delta\lambda - g(z), \\ U_t = V\lambda_t. \end{cases} \tag{19.26}$$

Finally, we can eliminate U. Differentiating the first equation in (19.26) with respect to t and using the second equation in (19.26) we obtain

$$\begin{aligned} V_{tt} &= \Delta U_t - 2\sum_{j=1}^{n}(V_{jt}\lambda_j + V_j\lambda_{jt}) - V_t\Delta\lambda - V\Delta\lambda_t \\ &= \Delta(V\lambda_t) - 2\sum_{j=1}^{n}(V_{jt}\lambda_j) - 2\sum_{j=1}^{n}V_j\lambda_{jt} - V_t\Delta\lambda - V\Delta\lambda_t \\ &= \lambda_t\Delta V - 2\sum_{j=1}^{n}\lambda_j V_{jt} - V_t\Delta\lambda. \end{aligned} \tag{19.27}$$

The Cauchy data for (19.27) is located on the plane $\{t = 0\}$ and is given by

$$V(z, 0) = 0, \tag{19.28}$$

since $u^*_\lambda(z, \lambda) = 0$ on S^λ for all λ. (Recall that $u^* \equiv f(z)$ on S^λ by (19.18).) The "normal" derivative of V, $V_t(z, 0)$, is found from (19.26) by setting $t = 0$, using $V(z, 0) = V_j(z, 0) = 0$ and that $U(z, 0) = u^*(z, \lambda(z, 0)) = u^*|_{S^\lambda} = f(z)$ by (19.25). Note that from (19.26) we also have $U_t(z, 0) = V(z, 0)\lambda_t = 0$. Thus, the equivalent problem (19.14) becomes in the case of the Laplacian

$$\begin{cases} V_{tt} &= \lambda_t\Delta V - 2\sum\limits_{j=1}^{n}\lambda_j V_{jt} - V_t\Delta\lambda, \\ V &\equiv t(\Delta f(z) - g(z)) \text{ on } Z_t = \left\{(z, t) \in \mathbb{C}^{n+1} : t = 0\right\}. \end{cases} \tag{19.29}$$

EXAMPLE 19.7. We present the problem of finding the Schwarz potential of the "sphere"

$$\Gamma = \left\{z \in \mathbb{C}^n : \varphi(z) = \sum_1^n z_j^2 - 1 = 0\right\},$$

as a series of exercises. For the sake of clarity we take $n \geq 3$. Thus, consider (19.18) for $g = 0$ and $f = \frac{1}{2}\sum\limits_1^n z_j^2$. As we know from Ch. 14, the solution is

$$u(z) = \frac{n}{2(n-2)} - \frac{\left(\sum\limits_1^n z_j^2\right)^{(2-n)/2}}{n-2}.$$

EXERCISE 19.1. Verify that

$$u^*(z, \lambda) = \frac{n(1+\lambda)}{2(n-2)} - \frac{(1+\lambda)^{n/2}\left(\sum\limits_1^n z_j^2\right)^{(2-n)/2}}{n-2} \tag{19.30}$$

is the solution of the corresponding problem (19.19) on

$$S^\lambda : \left\{(z, \lambda) \in \mathbb{C}^{n+1} : \sum_1^n z_j^2 - 1 = \lambda\right\}.$$

Hence, (cf. (19.20)) conclude that

$$u^*_\lambda = V = \frac{n}{2(n-2)}\left(1 - \left(\sum_1^n z_j^2\right)^{(2-n)/2}(1+\lambda)^{(n/2)-1}\right). \tag{19.31}$$

EXERCISE 19.2. From (19.21) show that the λ-function is equal to

$$\lambda = (1+4t)^{-1}\left(\sum_1^n z_j^2\right) - 1. \tag{19.32}$$

Therefore, by (19.32)

$$V(z,t) = u_\lambda^*(z,\lambda(z,t)) = \frac{n}{2(n-2)}\left(1-(1+4t)^{(2-n)/2}\right). \tag{19.33}$$

EXERCISE 19.3. Check that the function (19.33) satisfies (19.29) with initial conditions $V(z,0)=0$, $V_t(z,0) = \Delta\left(\frac{1}{2}\sum_1^n z_j^2\right) = n$.

EXERCISE 19.4. Finally, show (as required) that

$$u(z) = U(z,t(z)) = \frac{n}{2(n-2)} - \frac{\left(\sum_1^n z_j^2\right)^{(2-n)/2}}{n-2}. \tag{19.34}$$

Hint: Use (19.15) to obtain

$$\begin{aligned} U(z,t) &= \frac{1}{2}\left(\sum_1^n z_j^2\right) + \int_0^t V(z,s)\lambda_s(z,s)\,ds \\ &= \frac{\left(\sum_1^n z_j^2\right)}{2(n-2)}\left[\frac{n}{1+4t} - \frac{2}{(1+4t)^{n/2}}\right], \end{aligned} \tag{19.35}$$

and from (19.32) we have

$$t(z) = \frac{1}{4}\left(\sum_1^n z_j^2 - 1\right). \tag{19.36}$$

Observe that the isotropic cone $\Gamma_0 = \left\{z\in\mathbb{C}^n : \sum_1^n z_j^2 = 1\right\}$ appears in (19.34) as a singularity set, not because of any singularities of the function $t(z)$ — which is entire — (see (19.36)), but due to singularity of V and, consequently, of an "outer" function U, on a plane $t=-1/4$ "far away" from the initial plane $\{t=0\}$.

At first glance it seems unclear why the solution $u=u_\Gamma$ should have singularities at all, since the initial variety $\Gamma = \left\{z\in\mathbb{C}^n : \sum_1^n z_j^2 - 1 = 0\right\}$ does not have characteristic (with respect to $\Delta = \sum_1^n \partial^2/\partial z_j^2$) points. However, if we consider the "projectivization" of Γ, i.e., imbed Γ into the projective space $\mathbb{CP}^n$ by introducing the projective coordinates $\xi_i = z_i\eta$, $i=1,\dots,n$ and η, where $(\xi_i,\eta)\subset\mathbb{C}^{n+1}\setminus\{0\}$, we immediately observe that so imbedded $\widehat{\Gamma}\subset\mathbb{CP}^n$, $\widehat{\Gamma} = \left\{(\xi,\eta)\in\mathbb{CP}^n := \sum_1^n \xi_i^2 - \eta^2 =: \hat{\varphi} = 0\right\}$ has a whole "circle" of characteristic points — the intersection of $\widehat{\Gamma}$, with the plane "at infinity" $\{\eta=0\}$. Indeed, in projective coordinates the Laplacian becomes $\eta^2 \sum_1^n \partial^2/\partial\xi_i^2$ and hence

$$\begin{aligned}\widehat{\Gamma}_{ch} &= \left\{(\xi,\eta) : \sum_1^n \xi_i^2 - \eta^2 = 0 \text{ and } 4\eta^2\left(\sum_1^n \xi_i^2\right) = 0\right\} \\ &= \left\{(\xi,\eta) : \sum_1^n \xi_i^2 = 0 \text{ and } \eta = 0\right\}.\end{aligned} \tag{19.37}$$

Note that (19.37) is also an intersection of $\widehat{\Gamma}_0$, the "projectivized" isotropic cone Γ_0, with the same plane $\{\eta = 0\}$ "at infinity".

EXERCISE 19.5. Verify that at all points of $\widehat{\Gamma}_{ch}$, $\widehat{\Gamma}_0$ is tangent to $\widehat{\Gamma}$.

Moreover, the initial value problem (19.21) defining the λ-function becomes (under "projectivization")

$$\begin{cases} \Lambda_t = -\eta^2 \sum_1^n \Lambda_j^2, & (\Lambda_j := \partial/\partial\xi_j), \\ \Lambda(\xi,\eta,0) = \sum_1^n \xi_i^2 - \eta^2 \end{cases} \tag{19.38}$$

whose solution is easily found (cf. (19.32))

$$\Lambda(\xi,\eta,t) = \left(1 + 4\eta^2 t\right)^{-1}\left(\sum_1^n \xi_i^2\right) - \eta^2. \tag{19.39}$$

Unfortunately, (19.39) shows at once that all points of

$$\widehat{\Gamma}_{ch} = \left\{(\xi,\eta) \in \mathbb{CP}^n : \sum_1^n \xi_i^2 = \eta = 0\right\}$$

are exceptional points, i.e., $\Lambda(\xi,\eta,t) = 0$ for all t when $(\xi,\eta) \in \widehat{\Gamma}_{ch}$, and therefore Leray's theory does not apply here. However, there is a slight extension of Leray's theorem (Theorem 19.5) essentially due to Dunau [**35**]. The result of Dunau states that if a surface Γ has the unique characteristic tangent Σ to Γ at a point $z^0 \in \Gamma_{ch}$, even when z^0 is exceptional, the conclusion of Leray's theorem still holds, i.e., the singularities of the solution of the Cauchy problem close to z^0 remain on Σ.

In our example, the isotropic cone $\widehat{\Gamma}_0$ is the unique characteristic tangent to $\widehat{\Gamma}$ at $\widehat{\Gamma}_{ch}$, and this explains why the singularities of u_Γ stay on Γ_0. Also, this example shows (cf. the Remark following Theorem 19.5) that as characteristic points move to infinity, the singularities of the solution to a Cauchy problem tend to become unbounded singularities, even if they are still located on the characteristic tangent Σ^∞ tangent to the initial variety Γ at infinity. (Σ^∞ may also be called the asymptotic characteristic of Γ.)

EXAMPLE 19.8. Let $n = 2$. In what follows, we present the "visually easier" (cf. Figure 1) case of the Schwarz potential of an ellipse as a series of exercises. Let us denote the ellipse by

$$\Gamma := \left\{\frac{x^2}{a^2} + \frac{y^2}{b^2} - 1 = 0, a > b > 0\right\},$$

with $(\pm c, 0) : c^2 = a^2 - b^2$ the foci, and

$$\widehat{\Gamma} := \left\{ (X, Y) \in \mathbb{C}^2 : \frac{X^2}{a^2} + \frac{Y^2}{b^2} - 1 = 0, a > b > 0 \right\},$$

is the complexified ellipse. Also, $\mathcal{L} = \Delta$.

EXERCISE 19.6. Show that $\widehat{\Gamma}_{ch} = \left\{ \left(\pm \frac{a^2}{c}, \pm \frac{b^2}{c} i \right) \right\}$ consists of 4 points.

EXERCISE 19.7. Check that the solution of the initial value problem (19.9)

$$\begin{cases} \lambda_t = -\left(\frac{\partial \lambda}{\partial X}\right)^2 - \left(\frac{\partial \lambda}{\partial Y}\right)^2, \\ \lambda(X, Y, 0) = \frac{X^2}{a^2} + \frac{Y^2}{b^2} - 1, \quad X, Y \in \mathbb{C} \end{cases}$$

is

$$\lambda(x, y, t) = \frac{X^2}{a^2 + 4t} + \frac{Y^2}{b^2 + 4t} - 1. \tag{19.40}$$

The solution u of the Cauchy problem for the Laplacian with entire data $f(X, Y)$ may have singularities according to (19.16) at the singularities of the implicit function $t(X, Y)$ defined from (19.40) and $\lambda_t = 0$ and of course at the two poles of $\lambda(X, Y, t)$ at $t_1 = -\frac{a^2}{4}$, $t_2 = -\frac{b^2}{4}$.

EXERCISE 19.8. Show that at $t_1 = -\frac{a^2}{4}$, $\lambda(X, Y, t) = 0$ yields $X = 0$, $Y = \pm ic$ while at $t_2 = -\frac{b^2}{4}$, $\lambda(X, Y, t) = 0$ produces points $Y = 0, X = \pm c$.

EXERCISE 19.9. Check that the Leray developable Σ consists of four complex lines tangent to $\widehat{\Gamma}$ at respective points of $\widehat{\Gamma}_{ch}$.

$$X \pm iY = c \text{ at } \left(\frac{a^2}{c}, \pm \frac{b^2}{c} i \right),$$

$$X \pm iY = -c \text{ at } \left(-\frac{a^2}{c}, \pm \frac{b^2}{c} i \right).$$

Note that the singularities of u caused by poles of $\lambda(X, Y, t)$ at points $(0, \pm ic)$ and $(\pm c, 0)$ already belong to Σ hence *all* the singularities of *any* Cauchy problem with entire data are indeed confined to the Leray developable Σ in the case of the ellipse. Moreover, this remains true if lower order terms (with entire coefficients) are included in the differential operator (with the Laplacian still as the principal part). The poles of λ do not produce any new singularities. This miracle is addressed in the next chapter in all dimensions.

Observe that $\Sigma \cap \mathbb{R}^2 = \{(\pm c, 0)\}$, thus in full compliance with the earlier two dimensional results (see Corollary 13.1 and Example 13.1).

Finally, let us consider an example of an exceptional point where Leray's Principle fails altogether. For the sake of simplicity and clarity, we again restrict ourselves to the "real" picture. The following example is essentially Leray's original example.

EXAMPLE 19.9. Consider the equation

$$x_1 u_1 + 2x_2 u_2 = 0, \quad (u_i = \partial u / \partial x_i). \tag{19.41}$$

Note that, formally, every function of the form $u(x_1, x_2) = f\left(x_2/x_1^2\right)$, where f is real-analytic, satisfies (19.41). Choosing $f(w) = \frac{1}{w - \tau}$, $\tau \in \mathbb{R}$ say, we get a solution that is singular along the parabola $\tau x_1^2 = x_2$ and nowhere else. Of course,

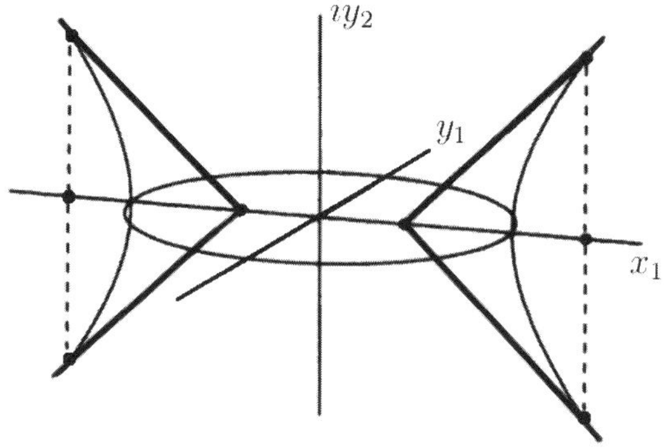

FIGURE 1. A three-dimensional slice of a complexified ellipse: The Schwarz potential of the ellipse has two algebraic singularities located at the foci $F_1 = (-c, 0)$ and $F_2 = (c, 0)$. Leray's principle reveals the true source of these singularities. The complexified ellipse has four characteristic points located at $\left(\pm\frac{a^2}{c}, \pm\frac{b^2}{c}i\right)$. The principal bicharacteristics emanating from those four points meet the real space precisely at the foci $(\pm c, 0)$.

all the curves $\sigma_\tau = \{(x_1, x_2) : \tau x_1^2 - x_2 = 0\}$ are everywhere characteristic (with respect to (19.41)) as they must be (including the limiting cases $\{x_i = 0\}$, $i = 1, 2$). Now consider a Cauchy problem for (19.41) with data on the non-singular curve $\Gamma := \{(x_1, x_2) : x_1^3 - x_2 = 0\}$. Its only characteristic point is at the origin. If we impose Cauchy data on Γ by

$$u\left(x_1, x_1^3\right) = f\left(x_1\right),$$

where f is a given function analytic near $x_1 = 0$, we get the solution $u = f\left(x_2/x_1^2\right)$. Now, taking $f(w) = (w - \tau)^{-1}$ and letting τ vary, we see that the singularities of u may occur along any of the characteristic parabolas tangent to Γ at the origin. Thus, in this case, different choices of regular Cauchy data near $(0, 0)$ lead to the singularities taking quite different, roughly speaking, unpredictable trajectories. This is only possible because of the degeneracy of the λ-function at $(0, 0)$; i.e., because $(0, 0)$ is in fact an exceptional point for (19.41) (cf. the discussion preceding Theorem 19.5). Let us indeed check that $(0, 0)$ is exceptional. To do this, we first have to solve (19.9) near Γ and then check that the equation $\lambda(x, t) = 0$ is not solvable in the neighborhood of $x^0 = (0, 0)$. Equation (19.9) has in our case the following form

$$\begin{cases} \lambda_t + x_1\lambda_1 + 2x_2\lambda_2 = 0 \quad (\lambda_i = \partial\lambda/\partial x_i), \\ \lambda(x, 0) = x_1^3 - x^2. \end{cases} \tag{19.42}$$

Writing a corresponding Hamilton-Jacobi system (cf. [**93**, Chapter I]), we obtain (introducing, as usual, the parametrization $x_1 = s_1$, $x_2 = s_2$ of the initial plane $\{t = 0\}$ in $\mathbb{R}^3$):

$$(19.43)\qquad \begin{cases} \dfrac{dx_1}{d\tau} = x_1, & x_1(0) = s_1, \\ \dfrac{dx_2}{d\tau} = 2x_2, & x_2(0) = s_2, \\ \dfrac{dt}{d\tau} = 1, & t(0) = 0, \\ \dfrac{d\lambda}{d\tau} = 0, & \lambda(0) = s_1^3 - s_2. \end{cases}$$

Solving (19.43) and eliminating s_1, s_2 we find

$$\lambda(x, t) = x_1^3 e^{-3t} - x_2 e^{-2t}.$$

Thus, $\lambda(0, 0, t) \equiv 0$, i.e., $(0, 0)$ is indeed an exceptional point.

Notes

Theorem 19.1 is due to Goursat. We follow the presentation from Hörmander's book [**85**]. The beautiful Theorem 19.2 is due to J. Hadamard's students De Lassus and LeRoux and can be found in Hadamard's treatise [**72**]. The converse was proven much later by Tsuno [**193**]. Section 2 is based on Leray's classical memoir [**126**] that is not an easy reading. We've tried to illustrate it with examples while presenting the essense of the beautiful and ingenious ideas of Leray. To that end we have mostly followed the exposition in G. Johnsson's Ph.D thesis (cf. [**94, 95**]). Hamada's extension of Leray's Principle can be found in [**73**]. Dunau's theorem extending it is in his paper [**35**]. Further references and examples can be found in G. Johnsson's scholarly treatise of the topic in [**94, 95**]. In our exposition of the three examples at the end of the chapter we also follow Johnsson's work [**94**].

CHAPTER 20

Global Propagation of Singularities in $\mathbb{C}^n$

1. Global propagation of singularities and Persson equations

Since as before our goal is to merely illustrate the main ideas behind Leray's theory rather than present the results in maximal generality we shall restrict our considerations and mainly focus on particular examples.

The uniformization (19.16) suggests two sources of singularities of solutions of the Cauchy problem.

(i) Singularities of the implicit function $t(z)$ calculated from the λ-function (19.9) and
(ii) singularities of the "outer" function $U(z,t)$ computed by calculating singularities of the solution V to (19.14) and then using (19.15).

From now on, we restrict our considerations to the "Laplace operator" $\Delta := \sum_1^n \partial^2/\partial z_j^2$. In that case, the equation for the λ-function (19.9) becomes (here and throughout this chapter, λ_j denotes $\frac{\partial \lambda}{\partial z_j}$ as in Chapter 19)

$$\begin{cases} \lambda_t + \sum_1^n \lambda_j^2 = 0, \\ \lambda(z,0) = \varphi(z), \end{cases} \tag{20.1}$$

where $\varphi(z)$ is the defining function of the initial variety Γ. In some special cases one can calculate λ explicitly (cf., e.g., (19.32)). Following G. Johnsson [**94**], we shall here take up two examples.

EXAMPLE 20.1. The complex hypersurface

$$\Gamma = \left\{ z \in \mathbb{C}^n : \sum_1^n z_i^2/a_i^2 - 1 = 0; a_1 > a_2 > \cdots > a_n > 0 \right\}$$

is a "complexified" heterogeneous ellipsoid ($\Gamma_{\mathbb{R}} = \Gamma \cap \mathbb{R}^n$ is an ordinary ellipsoid), and the solution of (20.1) is

$$\lambda(z,t) = \sum_{i=1}^n \frac{z_i^2}{a_i^2 + 4t} - 1. \tag{20.2}$$

EXERCISE 20.1. Check that (20.2) above as well as (20.3) below each solve (20.1) for $L = \Delta$.

EXAMPLE 20.2. $\Gamma = \left\{ z \in \mathbb{C}^n : \sum_1^{n-1} z_j^2 - z_n = 0 \right\}$ is a "complexified" paraboloid of revolution. Then,

$$\lambda(z,t) = \sum_1^{n-1} \frac{z_j^2}{1+4t} - z_n - t. \tag{20.3}$$

We indicate the main steps in calculating (20.3); (20.2) is obtained in a similar manner. The Hamilton-Jacobi system corresponding to (20.1) in the case of a paraboloid (cf. [**93**, Chapter I]) becomes

$$\begin{cases} \dfrac{dz_i}{d\tau} = 2\lambda_i, z_i(0) = s_i, i = 1, \ldots, n, \\ \dfrac{d\lambda_i}{d\tau} = 0, \lambda_i(0) = 2z_i, i = 1, \ldots, n-1, \lambda_n(0) = -1, \\ \dfrac{dt}{d\tau} = 1, \\ \dfrac{d\lambda}{d\tau} = \lambda_t + 2\sum_{j=1}^{n} \lambda_j^2, \lambda(z,0) = \sum_{i=1}^{n-1} z_i^2 - z_n. \end{cases} \tag{20.4}$$

Solving (20.4) for each index i, we obtain

$$\begin{cases} z_i = 2\lambda_i t + s_i, i = 1, \ldots, n, \\ \lambda = t\left(4\sum_1^{n-1} s_i^2 + 1\right) + \sum_1^{n-1} s_i^2 - s_n, \\ \lambda_i = 2s_i, i = 1, \ldots, n-1, \\ \lambda_n = -1. \end{cases}$$

And finally, eliminating parameters s_i, we arrive at (20.3).

Note, (and this is true in general as was shown by Johnsson [**94**], for an arbitrary quadratic surface) that $\lambda(z,t)$ in Examples 20.1 and 20.2 (also cf. (19.32)) has a particularly simple form

$$\lambda(z,t) = \frac{Q(z,t)}{r(t)},$$

where Q, r are polynomials; and while the degree of Q with respect to z is ≤ 2, the denominator r is a polynomial of one variable t. Now recall that the transformed Cauchy problem (19.14) in our situation ($\mathcal{L} = \Delta = \sum_1^n \partial^2/\partial z_j^2$) reduces to

$$\begin{cases} V_{tt} = \lambda_t \Delta V - 2\sum_{j=1}^{n} \lambda_j V_{jt} - V_t \Delta\lambda, \\ V \equiv t\Delta f \text{ on } \left\{(z,t) \in \mathbb{C}^{n+1} : t = 0\right\}, \end{cases} \tag{20.5}$$

(cf. (19.29), where we have taken $g = 0$ and the initial data $f(z)$ to be an entire function). Since in the above examples (and, in general, for all quadratic surfaces) λ_t is a quadratic polynomial in z and the λ_j are linear polynomials in z, the global behavior of (20.5) is covered by the following theorem of Miyake [**143**] and Persson [**150**].

THEOREM 20.1. *Consider the following non-characteristic Cauchy problem in* $\mathbb{C}^{n+1}$

$$\text{(20.6)} \quad \begin{cases} \dfrac{\partial^m u}{\partial t^m} + \sum\limits_{k=0}^{m-1} \sum\limits_{|\alpha|=m-k} a_\alpha(z,t) \dfrac{\partial^k}{\partial t^k} (\partial_z^\alpha u) + \sum\limits_{|\beta| \le m-1} b_\beta(z,t)\, \partial_{z,t}^\beta u = g(z,t); \\ \dfrac{\partial^k}{\partial t^k}\, u(z,t)|_{t=0} = f_k(z), 0 \le k \le m-1, \end{cases}$$

where $f_k(z)$ *are entire and the coefficients* $a_\alpha(z,t)$ *are polynomials in* z *such that*

$$\text{(20.7)} \qquad a_\alpha(z,t) = \sum_{|\gamma| \le |\alpha|} c_\gamma(t)\, z^\alpha,$$

where $c_\gamma(t)$ *as well as* $b_\beta(z,t)$ *and* $g(z,t)$ *are holomorphic in* $\mathbb{C}_a^n \times \Omega$, Ω *being a domain in* $\mathbb{C}_t$. *Then the solution* $u(z,t)$ *extends holomorphically (i.e., along any arc) to* $\mathbb{C}_z^n \times \Omega$.

REMARK 20.1. The equations (20.6) satisfying condition (20.7) are called Persson equations. J. Persson [**150**] proved Theorem 20.1 for the case of entire coefficients; Miyake [**143**] proved a general theorem comprising even Goursat problems. A very nice proof of Theorem 20.1 based on ideas of Bony, Schapira, and Zerner (cf. Ch. 8) was given by Johnsson in [**95**].

We postpone the discussion of the ideas behind Johnsson's proof of Theorem 20.1 while concentrating on applying this result to treat general Cauchy problems for Examples 20.1 and 20.2. Since for those examples, $\lambda = Q(z,t)/r(t)$, where the degree of Q in z is at most two (cf. (20.2), (20.3)), the equation (20.5) is a Persson equation and Theorem 20.1 can be applied. This means that the solution $V(z,t)$ and hence the "outer" function $U(z,t)$ (cf. (19.15)) extend holomorphically to $\mathbb{C}_z^n \times \Omega$, where $\Omega = \mathbb{C}_t \setminus \{\alpha_j\}$, α_j's being the poles of $\lambda(z,t)$ in $\mathbb{C}_t$. Hence the solution $u(z) = U(z,t(z))$ of the original problem (we recall that the notation $u \equiv f$ on Γ indicates that u and its gradient coincide with that of the data function f)

$$\text{(20.8)} \qquad \begin{cases} \Delta u &= 0, \\ u &\equiv f \text{ on } \Gamma \end{cases}$$

can have singularities in the set $\left\{ \bigcup\limits_j t^{-1}(\alpha_j) \right\}$ which can be attributed to the "outer" function $U(z,t)$ in addition to singularities of the implicit function $t(z)$ defined by $\lambda(z,t) = 0$.

Now let us apply this to our two examples.

EXAMPLE 20.3 (continuation of Example 20.1). As is seen from (20.2), $\lambda(z,t)$ has poles at $\alpha_j = -a_j^2/4$, $j = 1, \ldots, n$, and therefore the set of singularities of the "outer" function U belongs to the union of so-called "focal caustics"

$$S_U = \bigcup_{j=1}^n \left\{ z_j = 0, \sum_{i:i\ne j} \frac{z_i^2}{a_i^2 - a_j^2} = 1 \right\}.$$

(The reason for this name is very simple: the family $\Gamma_t := \{z : \lambda(z,t) = 0\}$ is a family of quadrics confocal with Γ, and so S_U is the union of intersections of surfaces of that family with coordinate planes.) However, for this particular case a miracle occurs. S_U turns out to be nothing else but the intersection of the Leray characteristic tangent (19.17) with coordinate planes, i.e., S_U is in fact the intersection of the singularities of the implicit function $t(z) : \lambda(z, t(z)) = 0$ with coordinate planes. To see this, we introduce for the sake of convenience a new parameter $s = 4t$. Then, the family $\{\Gamma_t\} = \{\Gamma_s\}$ can be described as satisfying

$$F(z, s) := \varphi(z, s)\psi(s) = 0, \tag{20.9}$$

where $\varphi(z, s) := \sum_{i=1}^{n} z_i^2 / \left(a_i^2 + s\right) - 1$, $\psi(s) := \prod_{i=1}^{n} \left(a_i^2 + s\right)$. The singularity set of the implicit function $t(z) = (1/4)s(z)$ lies on the envelope of the family (cf. (19.17)) which is defined by (20.9), together with

$$\varphi_s'(z, s)\,\psi(s) + \varphi(z, s)\,\psi'(s) = F_s'(z, s) = 0. \tag{20.10}$$

Now take, for instance, a "caustic"

$$\mathcal{C}_1 := \left\{ z \in \mathbb{C}^n : z_n = 0, \sum_1^{n-1} \frac{z_i^2}{a_i^2 - a_n^2} = 1 \right\}.$$

Equation (20.9) is satisfied since $\varphi|_{\mathcal{C}_1} = 0$, while (20.10) holds, because the value of s corresponding to $\mathcal{C}_1$ is $-a_n^2$, a root of the polynomial ψ, and also $\varphi\left(z, -a_n^2\right)$ vanishes on $\mathcal{C}_1$. (Of course, strictly speaking, $\varphi\left(z, -a_n^2\right)$ is not defined on $\mathcal{C}_1$, but it extends as zero to $\mathcal{C}_1$ by continuity.) Thus, the "outer" function U does not bring in any new singularities and all possible singularities of solutions of a Cauchy problem lie on the "Leray characteristic tangent" Σ defined by the equations (19.17), where $\lambda(z, t)$ is determined by (20.1).

Since Σ is generated by "principal bicharacteristics," which in the case of the Laplace operator are complex lines (cf. Example 19.4), it is a ruled algebraic surface. Moreover, it is not difficult to see that, for $n > 2$, Σ is an irreducible variety (cf. [**94**]). Thus, following Johnsson, we can strengthen Corollary 16.5 as follows.

THEOREM 20.2 ([**94**]). *All solutions u of the Cauchy problem*

$$\begin{cases} \Delta u &= 0, \\ u &\equiv f \text{ on } \Gamma = \left\{ z \in \mathbb{C}^n : \sum_1^n z_1^2 / a_i^2 = 1 \right\} \end{cases} \tag{20.11}$$

with entire data f extend holomorphically along all paths in $\mathbb{C}^n$, avoiding the "characteristic tangent" Σ.

REMARK 20.2. As follows from the above discussion,

$$\Sigma \cap \mathbb{R}^n = \bigcup_{j=1}^{n} \left\{ x \in \mathbb{R}^n : x_j = 0, \sum_{i:i \neq j} \frac{x_i^2}{a_i^2 - a_n^2} = 1 \right\},$$

i.e., the union of $(n-1)$ "focal" hyperboloids γ_j, $j = 1, \ldots, n-1$ (of k sheets, $k \leq n-2$) and a focal ellipsoid

$$\gamma_0 = \partial E = \left\{ x \in \mathbb{R}^n : x_n = 0, \sum_1^{n-1} \frac{x_i^2}{a_i^2 - a_n^2} = 1 \right\},$$

where, obviously,

$$E = \left\{ x \in \mathbb{R}^n : x_n = 0, \sum_{i=1}^{n-1} \frac{x_i^2}{a_i^2 - a_n^2} - 1 \leq 0 \right\}.$$

(Recall that we assume $a_1 > \cdots > a_n > 0$.) Thus, *a priori*, solutions of (20.11) with data on $\Gamma_{\mathbb{R}} = \Gamma \cap \mathbb{R}^n$ could exhibit singularities there, which is worse than what we obtained in Corollary 16.5. However, a closer look at the implicit function $t(x) = t(z)|_{\mathbb{R}^n}$ shows that as long as it is continued along a path in $\mathbb{R}^n$, it becomes multi-valued only when continued around γ_0, and therefore the only singularities which can actually occur while continuing solutions of (20.11) from $\Gamma_{\mathbb{R}}$ in $\mathbb{R}^n$ must all lie on the "focal ellipsoid" γ_0 exclusively, bringing the "real" version of Theorem 20.2 in complete correspondence with the earlier Corollary 16.5 for polynomial data. (To see that $t(x)$ extends with no trouble to all γ_j, $j = 1, \ldots, n-1$ and does not assume "forbidden values" $-a_j^2$, $j = 1, \ldots, n$ there, we simply observe that $t = t(x)$ is nothing other than an elliptic coordinate corresponding to the unique ellipsoid $\Gamma_{t,\mathbb{R}}$ of the family of quadrics confocal with $\Gamma_{\mathbb{R}}$ passing through x (cf. (20.1) and [**99**, Ch. VII, §4]. Since through all points of $\mathbb{R}^n \setminus E$ passes one and only one such ellipsoid, the analyticity and "continuability" of $t(x)$ to all of $\mathbb{R}^n \setminus \partial E$ follows.

However, if we extend a solution of (20.11) through $\mathbb{C}^n$, it may happen that arriving back to $\mathbb{R}^n$, we'll have gained a non-trivial period around some $\mathcal{C}_j$, $j = 1, \ldots, n-1$, and because of this, we may encounter a forbidden value $-a_k^2$, $k = 1, \ldots, n$. This effect (viewed, of course, in $\mathbb{C}^n$ context only) has been called by Johnsson a latent singularity, and is investigated in [**94**] in great detail.

EXAMPLE 20.4 (continuation of Example 20.2). From (20.3) we see that the only finite t-pole of $\lambda(z,t)$ is $\alpha = -1/4$. The singularity set of the "outer" function U caused by α is then the $(n-1)$-dimensional isotropic cone

$$\gamma = \left\{ z \in \mathbb{C}^n : z_n = 1/4, \sum_{1}^{n-1} z_j^2 = 0 \right\}$$

with the vertex at the "focus" $\mathcal{F} = (0, \ldots, 0, 1/4)$ of Γ. Singularities of the implicit function $t(z)$ are easily calculated from (20.3) by solving the quadratic equation $\lambda = 0$ for t. We find that

$$t(z) = \frac{-4z_n - 1 - \sqrt{(4z_n - 1)^2 + 16 \sum_{1}^{n-1} z_j^2}}{8}, \tag{20.12}$$

and hence $t(z)$ is holomorphic everywhere in $\mathbb{C}^n$ except for points on the isotropic cone $\Sigma = \left\{ z \in \mathbb{C}^n : \sum_{1}^{n-1} z_j^2 + (z_n - 1/4)^2 = 0 \right\}$ with the vertex at the focus $\mathcal{F}$. Note that $\gamma \subset \Sigma$. It is not hard to see that Σ is precisely the Leray characteristic tangent to the paraboloid Γ at finite characteristic points.

Indeed,

$$\Gamma_{ch} = \left\{ z \in \Gamma : 4 \sum_{j=1}^{n} z_j^2 + 1 = 0 \right\} = \left\{ z_n = -1/4, \sum_{j=1}^{n-1} z_j^2 = -1/4 \right\} = \Sigma \cap \Gamma,$$

and it is easy to check (an exercise for the reader) that Σ is tangent to Γ at Γ_{ch}. However, there is another source of singularities for the solution u of the Cauchy problem (20.8) coming from the multi-valuedness of the function (20.12) in $\mathbb{C}^n \setminus \Sigma$. Indeed, choosing a single-valued branch of $t(z)$ in (20.12), we find that it equals $-z_n$ on the "conical cylinder" $\Sigma^\infty := \left\{ z \in \mathbb{C}^n : \sum_1^{n-1} z_j^2 = 0 \right\}$. However, continuing $t(z)$ along a homotopically non-trivial path in $\mathbb{C}^n \setminus \Sigma$ and arriving back on Σ^∞, we find ourselves on the second branch of the square root, and hence so continued

$$t(z)|_{\Sigma^\infty} = \frac{-4z_n - 1 + (4z_n - 1)}{8} = -\frac{1}{4}$$

i.e., admits the "forbidden value" $-1/4$ where the "outer" function U may become singular. Therefore, solutions $u = U(z, t(z))$ of (20.8) can also become singular when $t(z)$ is continued from points of Σ^∞ "around" Σ on a homotopically nontrivial path in $\mathbb{C}^n \setminus \Sigma$. Thus, here we again have a new variety Σ^∞ as a carrier of singularities resulting from analytic continuation of solutions of the Cauchy problem from non-characteristic points of Γ along homotopically non-trivial paths around the Leray characteristic tangent Σ. Those are, again, latent singularities in Johnsson's terminology. (The fact that the singularity on Σ^∞ must be a latent singularity, i.e., be a result of analytic continuation along a certain path is not entirely unexpected; just note that no solution u of a Cauchy problem (20.11) can be immediately singular on Σ^∞ near the origin, for 0 is not a characteristic point of Γ.) Summarizing, we have the following result.

THEOREM 20.3 ([**94**]). *All solutions u of the Cauchy Problem*

$$\text{(20.13)} \qquad \begin{cases} \Delta u &= 0, \\ u &\equiv f \text{ on } \Gamma = \left\{ z_n \in \mathbb{C}^n : \sum_1^{n-1} z_j^2 - z_n = 0 \right\}, \end{cases}$$

with entire data f extend holomorphically along any path in $\mathbb{C}^n \setminus (\Sigma \cup \Sigma^\infty)$ and singularities may occur on Σ^∞ only if continuation of u takes place along a path not homotopic to zero in $\mathbb{C}^n \setminus \Sigma$. Here Σ and Σ^∞ are defined as above.

REMARK 20.3. (i) By embedding Γ into $\mathbb{CP}^n$, we see similarly to Example 19.6, that the so embedded surface

$$\widehat{\Gamma} = \left\{ (\xi, \eta) \in \mathbb{CP}^n : \sum_1^{n-1} \xi_i^2 - \eta\xi_n = 0 \right\}$$

has "at infinity" the whole cone of characteristic points

$$\widehat{\Gamma}_{ch}^\infty = \left\{ (\xi, \eta) : \eta = 0, \sum_1^{n-1} \xi_i^2 = 0 \right\}.$$

Moreover, similarly to Example 19.6, we find that those points are exceptional, hence Leray's theory does not apply. The characteristic variety $\left\{ (\xi, \eta) \in \mathbb{CP}^n : \sum_1^{n-1} \xi_i^2 = 0 \right\}$, which, as we have seen, serves as one of the carriers of singularities for solutions to the Cauchy problem, is not tangent to $\widehat{\Gamma}_{ch}$. Hence, even the Dunau strengthening of Leray's theorem does not apply here. The latter, of course, can be verified by direct computation: $\widehat{\Gamma}$ has no characteristic tangent at $\widehat{\Gamma}_{ch}$. Thus, here we have a crisp example of a singularity set ("strata" as christened by Sternin and Shatalov [**181**]) that appears not near Leray's tangent but as a consequence of continuing the solution around Leray's tangent along a homotopically non-trivial path. A nice justification of the Johnsson's term "latent singularity".

(ii) Let us again make a remark concerning "real" continuation of solutions u of (20.13) with data on $\Gamma_{\mathbb{R}} = \Gamma \cap \mathbb{R}^n$, i.e., continuation restricted to $\mathbb{R}^n$. As is easily seen,

$$\Sigma_{\mathbb{R}} = \left\{ x \in \mathbb{R}^n : \sum_{1}^{n-1} x_j^2 + (x_n - 1/4)^2 = 0 \right\} = \{(0, \ldots, 0, 1/4)\}$$

i.e., a focus, while $\Sigma_{\mathbb{R}}^{\infty} = \{x' := (x_1, x_2, ..., x_{n-1}) = 0\}$, the x_n-axis. Now, take the branch (20.12) for $t(x)$ (i.e., such that $t(0) = 0$) at any point on the paraboloid $\Gamma_{\mathbb{R}}$ (all points of $\Gamma_{\mathbb{R}}$ are, of course, non-characteristic) and continue $t(x)$ to an arbitrary point on the x_n-axis with $x_n < -1/4$. Then, even allowing our path to circle the ray ℓ^- of $\Sigma_{\mathbb{R}}^{\infty}$ with $x_n < -1/4$, we obviously end up with the same branch of $t(x)$ on ℓ^- without $t(x)$ ever assuming the "forbidden value" $-1/4$. Hence, the ray $\ell^- \subset \Sigma_{\mathbb{R}}^{\infty}$ can be dropped from the list of "real" singularities for solutions of (20.13) in $\mathbb{R}^n$.

On the other hand, if we are trying to continue $t(x)$ from the paraboloid $\Gamma_{\mathbb{R}}$ to the points on the closed ray $\ell^+ = \Sigma_{\mathbb{R}}^{\infty} \setminus \ell^-$, in order to stay away from the value $(-1/4)$ we must remain on the second branch of (20.12), the one for which $t(0) = -1/4$. However, any solution u of the Cauchy problem (20.13) near an arbitrary point $x^0 \in \Gamma_{\mathbb{R}}$ can be obtained by continuing it from another chosen point, say the origin, while avoiding the second (after the origin) encounter with the x_n-axis. (This can be done by successively applying the Cauchy-Kovalevskaya theorem to a finite chain of small balls stretched from the origin to x^0 along the surface $\Gamma_{\mathbb{R}}$ with centers on $\Gamma_{\mathbb{R}}$ without ever crossing the x_n-axis, except for the first ball centered at the origin.) Therefore, we must choose for any solution u of (20.13) the same branch ($t(0) = 0$) of $t(x)$ in (20.12). But that branch assumes the value $-1/4$ on the whole closed ray $\ell^+ := \{x' = 0, x_n \geq 1/4\}$. Thus, solutions of the Cauchy problem (20.13) may (and will) become singular on ℓ^+. So, the "real" singularities must all lie on $\ell^+ = \{x' = 0, x_n \geq 1/4\}$ and can actually occur at any point on that ray. Thereby, unlike the previous example, some of the latent singularities here remain, even if holomorphic continuation is restricted to $\mathbb{R}^n$.

Here is a heuristic illustration that might help the reader. Assume for simplicity that $n = 3$ (or even 2!). Then the interior of $\Gamma_{\mathbb{R}}$ (i.e., the convex component of the complement of $\Gamma_{\mathbb{R}}$) as is easily seen can be viewed as an increasing union of the interiors of ellipsoids E_n with axial symmetries, i.e., prolate spheroids, with focus F_1 at $x_3 = 1/4$ and the second focus on the x_3-axis running to $+\infty$. As we know from the previous example and the results in Ch. 13, the singularities of the solutions to the Cauchy problem (for $\mathcal{L} = \Delta$) posed on those ellipsoids might (and will, e.g., for the Schwarz potentials of ∂E_n) become singular on the "focal segment" joining F_1 and F_2. Since $[F_1, F_2] \to [F_1, \infty) = \{x_1 = x_2 = 0, 1/4 \leq x_3 < \infty\}$, this confirms our conclusion above. Incidentally, note that there are *no* singularities ever that occur in the non-convex part of the complement $\Omega_0 := \mathbb{R}^n \setminus \Gamma_{\mathbb{R}}$. This is the reason that Ω_0 is deemed a "null-quadrature domain", i.e. $\int_{\Omega_0} u = 0$ for all $u \in \operatorname{Harm}(\Omega_0) \cap L^1(\Omega_0)$; cf. Ch. 22 for more details.

(iii) Due to the fact that all considerations in Ch. 19 rest in their essential part on the principal part P_m of the differential operator, Example 19.6, Theorems 20.2 and 20.3 can be easily extended to linear operators $\mathcal{L} = \Delta+$ (lower order terms), where the coefficients in the lower order terms are assumed to be entire functions. This is done in G. Johnsson's thesis and we refer the reader to his paper [**94**] for a detailed account of these results.

(iv) Theorem 20.3 has also been proved by B. Sternin and V. Shatalov in [**181**] by their method of explicit integral representations of solution of Cauchy problems.

At this point we leave it to the reader to find further results and detailed discussions of the Cauchy problem on quadric surfaces for more general PDEs with constant coefficients in the principal part in G. Johnsson's treatise [**94**]. Johnsson also gives a detailed account of classification of the singularities that might occur where solutions are continued in $\mathbb{R}^n$ and, more generally, in $\mathbb{C}^n$.

We would like to conclude this section by indicating the ideas of Johnsson's proof of Persson's theorem (Theorem 20.1). We shall sketch the argument for one very simple example, referring the reader to Johnsson's paper [**95**] for the proof in the general case. Let us note here that although the following example can be treated directly, it serves as a very good model for all the features that appear in Johnsson's general construction. So, consider the operator $\mathcal{L} = \partial/\partial z_2 - z_1\partial/\partial z_1$. $\mathcal{L}$ certainly satisfies condition (20.7), and therefore is a Persson operator. First, observe that a vector $\xi = (\xi_1, \xi_2)$ at a point $(z_1, z_2) \in \mathbb{C}^2$ is characteristic with respect to $\mathcal{L}$ if and only if

$$\xi_2 - z_1\xi_1 = 0.$$

Thus, in particular,

$$|\xi_2| / |\xi_1| < 2|z_1| \tag{20.14}$$

always holds for characteristic vectors. Now consider a domain

$$\Omega(\rho) = \left\{z \in \mathbb{C}^2 : |z_1| < \rho\left(|z_2|\right)\right\},$$

where $\rho : \mathbb{R} \to \mathbb{R}$ is a non-increasing function, say of class C^1, satisfying

$$\text{(i)} \quad \rho(0) > 0 \qquad \text{and} \qquad \text{(ii)} \quad \rho\left(t_0\right) = 0 \text{ for some } t_0 > 0, \tag{20.15}$$

to be specified later. Then, $\Omega(\rho)$ is a smoothly bounded domain in $\mathbb{C}^2$ containing $(0, 0)$. Now let $\xi = (\xi_1, \xi_2)$ be a "complex normal vector" to the boundary

$$\partial\Omega_\rho = \left\{z \in \mathbb{C}^2 : \rho\left(|z_2|\right) - |z_1| =: \varphi\left(z_1, \bar{z}_1, z_2, \bar{z}_2\right) = 0\right\}$$

at point (z_1, z_2). Then

$$|\xi_2| / |\xi_1| = \left|\rho'\left(|z_2|\right)\right|. \tag{20.16}$$

Indeed, assuming first that neither z_1 nor $z_2 = 0$, we obtain, since $(\xi_1, \xi_2) = (\partial\varphi/\partial z_1, \partial\varphi/\partial z_2)$,

$$\xi_1 = -\frac{\overline{z_1}}{2|z_1|},$$

$$\xi_2 = \rho'\left(|z_2|\right)\frac{\overline{z_2}}{2|z_2|}$$

and (20.16) follows. Either of the remaining cases ($z_1 = 0$ or $z_2 = 0$) follows then by continuity. Now suppose for a moment (in violation of (20.15(ii)), that ρ satisfies

$$\rho'(t) = -2\rho, \qquad \text{i.e.,} \qquad \rho(t) = \rho_0 e^{-2t} \tag{20.17}$$

for, say, $t \geq 0$. Then, the complex normal $\xi = (\xi_1, \xi_2)$ at a point $(z_1, z_2) \in \partial\Omega(\rho)$ satisfies by (20.16) and the definition of $\Omega(\rho)$

$$|\xi_2| \,/\, |\xi_1| = |\rho'\left(|z_2|\right)| = 2\rho\left(|z_2|\right) = 2\,|z_1| > |z_1|,$$

and therefore, in view of (20.16), cannot be characteristic (cf. (20.14)) unless $z_1 = 0$. Thus, locally by Zerner's theorem, if the solution u of the Cauchy problem

$$\begin{cases} Lu &= 0, \\ u\left(z_1, 0\right) &= f\left(z_1\right), \quad f \text{ is entire,} \end{cases} \tag{20.18}$$

where holomorphic in $\Omega\left(\rho_0\right)$, it would extend holomorphically across every point of $\partial\Omega\left(\rho_0\right) \setminus \{z_1 = 0\}$. Then if we were able to apply the compactness argument (as in the proof of Corollary 5.1), we could extend u holomorphically to $\Omega\left(\rho\right)$, $\rho(t) = \left(\rho_0 + \epsilon\right) e^{-2t}$, $\epsilon > 0$ and be almost done. For the union over all positive ϵ of all such domains $\Omega(\rho)$, with $\rho = \left(\rho_0 + \epsilon\right) e^{-2t}$, covers all of $\mathbb{C}^2$ and therefore all solutions of (20.18) would be proven to extend holomorphically to $\mathbb{C}^2 \setminus \{z_1 = 0\}$. This remaining obstacle is easily removed by copying the proof of Corollary 5.1. In other words, cover $\mathbb{C}^2$ by concentric balls $B_r = \left\{z \in \mathbb{C}^2 : |z| < r\right\}$ observing that the only characteristic points on $\partial B_r = \left\{z \in \mathbb{C}^2 : z_1\bar{z}_1 + z_2\bar{z}_2 - r^2 = 0\right\}$ are points satisfying

$$\begin{cases} \overline{z_2} - |z_1|^2 = 0, \\ |z_1|^2 + |z_2|^2 = r^2, \end{cases}$$

which are "far away" from the "suspicious" coordinate plane $\{z_1 = 0\}$. Hence by Zerner's theorem, if u is holomorphic in B_r, it extends to be holomorphic in $B_{r+\epsilon}$ for some $\epsilon > 0$. Since by the Cauchy-Kovalevskaya theorem it is initially holomorphic in some B_{r_0}, $r_0 > 0$, u is finally extended to all of $\mathbb{C}^2$, i.e., all solutions of (20.18) are entire, in accordance with the Persson-Miyake theorem (Theorem 20.1). The only remaining gap in our argument is that we cannot by simply applying the Cauchy-Kovalevskaya theorem extend u to be "initially" holomorphic in $\Omega\left(\rho_0\right)$ for some initial $\rho_0 > 0$ and later apply compactness, since for so defined $\rho(t)$ domains $\Omega(\rho)$ are not bounded (the condition (20.15(ii)) is violated). This can be easily rectified though, by replacing ρ as in (20.17) by a "cut off" function ρ defined as follows

$$\rho(t) = \begin{cases} \rho_0 e^{-2t}, & 0 \leq t \leq T, \\ \sigma(t), & T \leq t \leq t_0, \\ 0, & t > t_0, \end{cases} \tag{20.19}$$

where σ is a decreasing C^1-function smoothly joined to $\rho_0 e^{-2t}$ at T, which "decays regularly and faster than $\rho_0 e^{-2t}$," or in more precise terms satisfying

$$\sigma'(t) \leq -2\sigma(t) \quad \text{and} \quad \sigma\left(t_0\right) = 0. \tag{20.20}$$

Such a σ will exist for arbitrary ρ_0, T, and t_0 as long as $T < t_0$, and we leave the construction of such $\sigma(t)$ to the reader as a simple calculus exercise. Note that (20.20) implies that $|\sigma'(t)| \geq \left|\left(\rho_0 e^{-2t}\right)'\right| = 2\rho_0 e^{-2t}$ and, therefore, boundaries of all domains $\Omega(\rho) = \left\{z \in \mathbb{C}^2 : |z_1| < \rho\left(|z_2|\right)\right\}$, where ρ is defined by (20.19), do not contain characteristic points other than those on the axis $\{z_1 = 0\}$. All $\Omega(\rho)$ are now bounded, and since we can choose t_0, ρ_0 to be (independently) arbitrarily large or small, we see that for some small ρ_0, t_0, the domains $\Omega(\rho)$ will be inside a neighborhood of the origin where the solution u of (20.10) is holomorphic by the Cauchy-Kovalevskaya theorem while the union of all such $\Omega_{t_0,\rho_0}(\rho)$ covers all of $\mathbb{C}^2$. Therefore, we can apply the argument above without any reservations and extend all solutions u first to $\mathbb{C}^2 \setminus \{z_1 = 0\}$ and thereafter to all of $\mathbb{C}^2$.

Closing this discussion, let us add that in order to get a better feeling for the above globalization scheme in relation to the Persson condition (20.7), we recommend as an exercise for the reader to investigate what parts of the argument break down if we replace the coefficient z_1 by z_1^2, and consider instead an operator $\mathcal{L}' = \partial/\partial z_2 - z_1^2\partial/\partial z_1$ (cf. the example in Sec. 5 of Ch. 1). The operator $\mathcal{L}$ obviously violates Persson's condition since the power 2 appearing in the $z_1^2\frac{\partial}{\partial z_1}$ term exceeds the order of the operator. So, the exercise for the reader is to check where exactly the construction of $\{\Omega_\rho\}$, as domains covering $\mathbb{C}^2$, fails.

Although $\mathcal{L}'$ violates Persson's condition, the following exercise shows that that operator still illustrates Leray's principle quite well.

EXERCISE 20.2. (i) Show that the initial plane $\Gamma = \{z_2 = 0\}$ has one characteristic point at infinity if we consider its embedding $\widehat{\Gamma}$ into $\mathbb{CP}^2$ (cf. Example 19.4).

(ii) Show that the everywhere characteristic (with respect to $\mathcal{L}'$) variety $\{z_2 = \frac{1}{z_1}\}$ is precisely the Leray tangent from a characteristic point at infinity. Thus, the singularities of all the solutions of the Cauchy Problem (13.3) with an entire data $f(z_1)$, that are, as is easily checked, equal to $f\left(\frac{z_1}{1-z_1z_2}\right)$ may only appear on that variety.

2. A note on characteristic surfaces for the Laplace operator

We conclude this chapter by posing the problem of investigating *everywhere characteristic surfaces* for the Laplace operator in $\mathbb{C}^n$, giving special attention to the case when such surfaces are algebraic, or even simpler, cones, i.e., given as the zero sets of homogeneous polynomials. This topic is motivated by the result of de Lassus and Le Roux discussed in Section 1 of Chapter 19, and it seems to fit naturally into the theme of the current chapter of global propagation of singularities, with characteristic surfaces being the carriers of singularities.

Recall that a holomorphic hypersurface $\Gamma = \{z \in \mathbb{C}^n : \varphi(z) = 0\}$ is *characteristic* with respect to $\Delta = \sum_{j=1}^n \frac{\partial^2}{\partial z_j^2}$ if the condition

$$\sum_{j=1}^{n} \left(\frac{\partial \varphi}{\partial z_j}\right)^2 = 0 \tag{20.21}$$

holds everywhere on Γ.

As we have observed on many occasions for $n = 2$, by introducing the "bicharacteristic" or "Study" coordinates $z = z_1 + iz_2$, $w = z_1 - iz_2$, we reduce (20.21) to the condition

$$\varphi_z \cdot \varphi_w,$$

and accordingly planes of the form $\{z = \text{const}\}$, or $\{w = \text{const}\}$ are the only characteristic surfaces in $\mathbb{C}^2$.

As we have seen earlier in this chapter in $n \geq 3$ dimensions there is an abundance of algebraic characteristic surfaces other than planes. The variety of characteristic surfaces increases dramatically with the dimension of the ambient space. The following simple result describes *all* characteristic homogeneous conoids in $\mathbb{C}^3$, showing that they are rather limited.

PROPOSITION 20.1. *Let $\phi = \phi(z_1, z_2, z_3)$ be an irreducible homogeneous polynomial such that the variety $\Gamma := \{z \in \mathbb{C}^3 : \varphi(z) = 0\}$ is everywhere characteristic (with respect to Δ). Then either*

$$\varphi(z_1, z_2, z_3) = \sum_{i=1}^{3} \alpha_i z_i,$$

where $\sum_{i=1}^{3} \alpha_i^2 = 0$, i.e., Γ is a characteristic plane, or

$$\varphi(z_1, z_2, z_3) = \sum_{i=1}^{3} z_i^2,$$

and accordingly Γ is an isotropic cone.

We present the proof as a series of exercises.

Choosing z where $\nabla\varphi(z) \neq 0$, we can solve $\varphi(z) = 0$ for, say z_3: $z_3 = \psi(z_1, z_2)$, so that

$$\varphi(z_1, z_2, \psi(z_1, z_2)) = 0. \tag{20.22}$$

EXERCISE 20.3. Show that

$$\psi_1^2 + \psi_2^2 = -1. \tag{20.23}$$

Hint: Use the fact that Γ is characteristic along with the two equations obtained from differentiating (20.22) with respect to each of the variables z_1 and z_2.

EXERCISE 20.4. Show that

$$z_1\psi_1 + z_2\psi_2 = \psi,$$

and conclude that ψ is homogeneous of order 1.
Hint: First show that

$$-z_1\varphi_3\psi_1 - z_2\varphi_3\psi_2 + z_3\psi_3 = n\varphi$$

using the fact that φ is homogeneous which implies that it satisfies Euler's equation

$$\sum_{i=1}^{3} z_j\varphi_j = n\varphi, \text{ where } n = \text{ degree of } \varphi.$$

Then use the equation of the surface $z_3 = \psi(z_1, z_2)$.

Since ψ is homogeneous of order 1, it can be expressed in polar coordinates as $\psi(r, \theta) = rf(\theta)$.

EXERCISE 20.5. Show that $f(\theta)$ satisfies the ordinary differential equation

$$(f')^2 + f^2 = 1. \tag{20.24}$$

Hint: Express ψ_1 and ψ_2 in terms of f' by differentiating in polar coordinates using $\frac{\partial}{\partial z_1} = \cos\theta\frac{\partial}{\partial r} - \frac{\sin\theta}{r}\frac{\partial}{\partial\theta}$, and $\frac{\partial}{\partial z_2} = \sin\theta\frac{\partial}{\partial r} + \frac{\cos\theta}{r}\frac{\partial}{\partial\theta}$. Then use the resulting expressions to obtain (20.24) from (20.23).

EXERCISE 20.6. Argue that there are two types of solutions of (20.24): either (I) $f = \pm 1$ is constant, in which case Γ must be an isotropic cone, or (II) $f = \alpha_1 \cos\theta + \alpha_2 \sin\theta, \alpha_1^2 + \alpha_2^2 = 1$ in which case Γ must be a plane.

Hint: Differentiate (20.24) in order to obtain a homogeneous ODE whose left hand side factors as a product of linear forms.

The scheme for reducing the number of variables and deriving the ordinary differential equation (20.24) will, for $n = 4$ dimensions and higher, instead lead to a partial differential equation. Thus, for higher dimensions the situation is more complicated, and one should expect more homogeneous characteristic surfaces in higher dimensions.

As a simple example of a characteristic cone that is not the isotropic cone, the six-dimensional hypersurface

$$p := (a_1x_1 + a_2x_2)^m + (a_3x_3 + a_4x_4)^m + (a_5x_5 + a_6x_6)^m = 0, \quad m \geq 2$$

is everywhere characteristic for the Laplacian, provided that $a_1^2 + a_2^2 = a_3^2 + a_4^2 = a_5^2 + a_6^2 = 0$, since this condition ensures that p satisfies $\nabla p \cdot \nabla p = 0$. Moreover, p is irreducible as the sum of three mth powers.

The following questions are open to the best of our knowledge.

Question I. Do there exist homogeneous irreducible surfaces besides characteristic planes and isotropic cones in dimensions 4, and 5?

Question II. Can one find a characteristic conoid (that is not the isotropic cone) such that the intersection with $\mathbb{R}^n$ is just the origin?

The latter question (if it is answered affirmatively) would serve as an example of a characteristic surface $\Gamma := \{\phi(z) = 0\}$ (with respect to the Laplacian) such that there are no harmonic functions u in $\mathbb{C}^n \setminus \Gamma$ satisfying $u = O(|\phi|^{-k}), k > 0$ near Γ, contrary to the isotropic cone which is a set of "polar" singularities for the Newtonian kernel in dimensions $n \geq 3$. Indeed, if u were such a function, u would be harmonic in $\mathbb{R}^n \setminus \{0\}$ (since the support of Δu is $\{0\}$) and satisfy $|u| = O(|x|^{-k})$ near the origin. Then by a corollary from the theorem of L. Schwartz concerning distributions with one-pont support [**63**, Sec. 4.6], we have, near the origin,

$$u(x) = \sum a_\alpha \partial^\alpha \left(|x|^{2-n}\right) + u_0,$$

where $a_\alpha \in \mathbb{C}$, the sum is finite, and u_0 is harmonic near 0. Note that $|x|^{2-n}$ is the Newtonian kernel in $\mathbb{R}^n$. But this implies that, in a neighborhood of the origin in $\mathbb{C}^n$, u is singular on the isotropic cone and nowhere else, a contradiction.

Recall from Section 1 of Chapter 19 that according to a theorem of Tsuno, for any piece Γ' of such hypothetical $\Gamma \setminus \{0\}$, there is a function harmonic in a neighborhood of Γ' and having a polar singularity on Γ'. However, the origin being a singular point of Γ does not allow to extend Tsuno's function to a neighborhood of the whole variety Γ. Thus, an affirmative answer to the Question (II) would show that Tsuno's theorem is sharp and cannot be extended even to isolated singular points of characteristic varieties.

Notes

The ingenious uniformization of solutions to holomorphic Cauchy problems for linear PDE has originated and been developed by J. Leray in a series of remarkable papers [**126**], [**124**], [**125**], [**127**], [**128**], [**129**] under the common title *Probléme de Cauchy* and later developed further by Leray jointly with L. Garding and T.

Kotake in [**62**]. Here we followed G. Johnsson's papers [**94, 95**] where he extended Leray's local theory to obtain very precise global continuation results for Cauchy problems for second order equations posed on general quadric surfaces. His method is based on global results of Persson [**150**] and M. Miyake [**143**] for linear equations with entire coefficients. G. Johnsson [**95**] has also given a very nice proof of Miyake-Persson theorem using the ideas of globalizing families of J. Bony and P. Schapira. In our presentation of the ideas of the proof we also followed Johnsson's argument [**95**]. A parallel approach to obtaining global results based on Leray's theory was pioneered in the 1990s by B. Sternin and V. Shatalov and their school (see [**181**], [**182**], [**166**], and the references cited there). It is based on global theory of asymptotic solutions developed by Maslov in the mid-1960s. The key tool is a certain integral transform reminiscent of Fourier-type integral operators. Leray's theory for, e.g., meromorphic or ramified non-entire Cauchy data was developed in works of Hamada [**73**] and C. Wagschal [**198**], [**74**]. An interested reader is also encouraged to confer with the later publications of these authors for further developments.

Many years ago H.S. Shapiro and the first author tried to launch the program of studying the character of singulairites of, say, the Schwarz potential of a surface depending on the precise information regarding how the Leray developable surface Σ hits $\mathbb{R}^n$. For example, the singularities at the foci of an ellipse, or at all points of the caustic of an oblate spheroid (Example 1.2, Ch. 1) are of a "square root type", and it seems to stem from the fact that at each point of the caustic or each focus there are precisely two principal bicharcteristics emanating from the charcteristic points on the complexified surface meeting in $\mathbb{R}^n$ at those points. Yet, each point on the focal segment of the prolate spheroid (Example 1.3, Ch. 1) is the intersection of infinitely many principal charcteristics resulting in an unbounded non-algebraic singularity. A similar observation holds for the sphere where the center is the intersection of infinitely many principal bicharacteristics lying on the "asymptotic" characteristic cone tangent to the projectivized surface at infinity. This research direction, though it appears to be promising and appealing, remains virtually unexplored.

CHAPTER 21

Quadrature Domains and Laplacian Growth

Consider the following moving boundary problem. We seek a smooth one-parameter family of domains, $\{\Omega_t\}$ in $\mathbb{R}^n$ such that the normal velocity v_n of the boundary $\Gamma_t := \partial\Omega_t$ is determined by the harmonic Green's function, $g(x,t)$, of Ω_t with singularity at a fixed position x_0 contained in each Ω_t, namely,

$$(21.1)\qquad \begin{cases} v_n|_{\Gamma_t} = -\partial_n g, \\ \Delta g = Q\delta_{x_0}, \text{ in } \Omega_t, \\ g|_{\partial\Omega_t} = 0. \end{cases}$$

This problem serves as a simplified model for the moving boundary of a viscous fluid. Let us explain informally the derivation of (21.1) as a fluid dynamic model. Imagine a region of incompressible viscous fluid within a porous medium surrounded by an inviscid fluid. Suppose there is a source or sink at position x_0 in the region occupied by viscous fluid. Averaging the Navier-Stokes equations over pores [**24**] leads to Darcy's law for the fluid velocity v in terms of the pressure g

$$(21.2)\qquad v = -\nabla g.$$

Incompressibility implies that the divergence of the velocity vanishes so that

$$\nabla \cdot v = -\Delta g = 0,$$

except at the source or sink. The pressure of the inviscid fluid is assumed constant (atmospheric), a reasonable assumption if the process is "quasi-static" (in which case pressure has time to continually equalize throughout the region of inviscid fluid).

If we neglect surface-tension[1] then the pressure matches at the interface, which gives a constant (say zero) boundary condition for g, so g is nothing more than the harmonic Green's function [**99**, Ch. IX, Sec. 3] with a singularity at x_0. The mathematical problem is thus, as summarized in (21.1), to track the evolution of a domain Ω_t whose boundary velocity is determined by the gradient of its own Green's function.

If the domains Ω_t are bounded, (21.1) actually produces a *shrinking* boundary. The term "Laplacian growth" refers to the case when Ω_t are unbounded and the sink is placed at infinity by prescribing asymptotics for ∇g - cf. Fig. 1. In the two-dimensional case, the model is associated with viscous fingering in a *Hele-Shaw cell.* Two sheets of glass are placed close together with a viscous fluid (say oil or glycerin) filling the void between them. A small hole is drilled in the center of the top sheet and an inviscid fluid (say water or air) is pumped in at a constant rate.

[1] Among the various simplifying assumptions in the derivation of the Laplacian growth model, the assumption of zero surface tension is the most controversial [**88**, **188**].

1. Dynamics of singularities of the Schwarz potential

Suppose $\partial\Omega_0$ is real-analytic. Then for some interval of time there exists a unique solution of (21.1) such that $\Gamma_t := \partial\Omega_t$ remains analytic and has analytic time-dependence (see [**189**]). Thus, Γ_t has a Schwarz potential as defined in (13.11) of Ch. 13.

THEOREM 21.1. *Suppose $\Omega_t \subset \mathbb{R}^n$ is a one-parameter family of domains with $\partial\Omega_t$ analytic and having analytic time-dependence. Then, letting $w(x,t)$ denote the Schwarz potential of Γ_t, Ω_t solves the Laplacian growth problem (21.1) if and only if*

$$\frac{\partial}{\partial t}w(x,t) = -ng(x,t). \tag{21.3}$$

We refer the reader to Ch. 13 for the statement of the Cauchy problem defining the Schwarz potential.

REMARK 21.1. In particular, singularities of the Schwarz potential in the domains Ω_t do not depend on time, except for one stationed at the sink.

REMARK 21.2. In the case of $n = 2$, the Schwarz potential $w(z,t)$ is related to Schwarz function $S(z,t)$ by the equation $S(z,t) = 2\partial_z w(z,t)$, thus we have

$$S_t = \frac{\partial}{\partial t}(2\partial_z w) = -4\partial_z g.$$

PROOF. It suffices to show that for each t, the functions $w_t(x,t)$ and $-ng(x,t)$ solve the same Cauchy problem. Then by the uniqueness part of the Cauchy-Kovalevskaya Theorem, they are identical. Since each function is harmonic, we just need to verify the Cauchy data, and this will follow from an application of the chain rule.

First, we will show that $w_t(x,t)|_{\Gamma_t} = 0$. Consider a point $x(t)$ which is on Γ_t at time t. The chain rule gives

$$\frac{d}{dt}w(x(t),t) = \nabla w(x(t),t)\cdot \dot{x}(t) + w_t(x(t),t), \tag{21.4}$$

where $\dot{x}$, etc. denotes the derivative with respect to time t.

On the other hand, by the first piece of Cauchy data in (13.11) defining w, we have

$$\frac{d}{dt}w(x(t),t) = \frac{d}{dt}\left(\frac{1}{2}|x(t)|^2\right) = x(t)\cdot\dot{x}(t). \tag{21.5}$$

By the second piece of Cauchy data in (13.11),

$$x(t)\cdot\dot{x}(t) = \nabla w(x(t),t)\cdot\dot{x}(t). \tag{21.6}$$

Combining (21.5) and (21.6) with equation (21.4) gives

$$w_t(x,t)|_{\Gamma_t} = 0. \tag{21.7}$$

We are done if we can show that $\nabla w_t|_{\Gamma_t} = -n\nabla g$. Fix x, and in a neighborhood of x let $T(x)$ be the value of time at which the boundary, Γ_t, of the growing domain passes x. So in this neighborhood we have $\Gamma_t = \{T(x) = t\}$, and $|\nabla T| = \frac{1}{v_n}$ where v_n is the normal velocity of Γ_t. Then by the Cauchy data defining the Schwarz potential,

$$\nabla w(x,t)\Big|_{t=T(x)} = x.$$

Taking the divergence of both sides and again using the chain rule:

$$\nabla w_t \cdot \nabla T + \Delta w = n. \tag{21.8}$$

Recall, $v_n = -\partial_n g$. Thus, $\nabla T = \frac{-\nabla g}{|\nabla g|^2}$. Substitution into equation (21.8) shows that $\nabla w_t \cdot \nabla g = -n|\nabla g|^2$ holds on Γ_t. By equation (21.7), $w_t|_{\Gamma_t} = 0$, which implies that ∇w_t and ∇g are parallel, and hence $\nabla w_t|_{\Gamma_t} = -n\nabla g$. □

2. Quadrature domains and Richardson's theorem

We say that Ω is a *quadrature domain* if it admits a quadrature formula for integration of harmonic functions u,

$$\int_\Omega u dV = \sum_{m=1}^{N} \sum_{|\alpha|=0}^{n_m} a_{m,\alpha} \partial^\alpha u(x_m), \tag{21.9}$$

where $x_1, x_2, .., x_N$ are points in Ω and $a_{m,\alpha}$ are real constants. In the planar case, where quadrature domains were first introduced, they were defined using analytic functions as the test class [**5**]. The following theorem describes quadrature domains in terms of the Schwarz potential, w, showing that w has a single-valued real-analytic continuation throughout Ω except for finitely many points, where Δw is supported as a distribution of finite order (see [**110**]).

THEOREM 21.2 ([**110**]). *Suppose that Ω is a connected domain in $\mathbb{R}^n$ with C^1 boundary. Then Ω is a quadrature domain with quadrature formula (21.9) if and only if the Schwarz potential of $\partial\Omega$ satisfies*

$$w(x) = n\left(H(x) - K(x)\right), \tag{21.10}$$

where $H(x)$ is harmonic in Ω and C^1 in $\overline{\Omega}$, and $K(x)$ is the result of applying the quadrature formula (21.9) to the fundamental solution $E_n(x,y) = c_n|x-y|^{2-n}$ (in its second argument).

For example, let $\Gamma := \{x \in \mathbb{R}^n : |x|^2 = r^2\}$ be a sphere of radius r. When $n = 2$, we leave it as an exercise to verify that

$$w(\zeta) = r^2\left(\log|\zeta| + 1/2 - \log(r)\right)$$

solves the Cauchy Problem (13.11), and in higher dimensions the Schwarz potential is

$$w(x) = \frac{n}{2(n-2)} r^2 - \frac{r^n}{(n-2)|x|^{n-2}}.$$

Let us check that this agrees with Theorem 21.2. The mean-value property for the ball B_r of radius r centered at zero gives the quadrature formula:

$$\int_{B_r} u dV = \mathrm{Vol}\left(B_r\right) u(0).$$

Applying this quadrature formula to $E_n(x,y)$ in its second argument gives

$$K(x) = \mathrm{Vol}(B_r)E_n(x) = \mathrm{Vol}(B_r)c_n|x|^{2-n} = r^n \mathrm{Vol}(B_1)c_n|x|^{2-n}$$
$$= \frac{r^n}{n(n-2)}|x|^{2-n},$$

since the constant c_n appearing in the fundamental solution is $\dfrac{1}{(n-2)\omega_n}$, where ω_n is the surface area of the unit sphere. Taking $H(x) = \dfrac{n}{2(n-2)}r^2$, we obtain the same formula (21.10).

Suppose Ω_0 is a quadrature domain, and Ω_t solves (21.1). Then Theorems 21.1 and 21.2 together imply that Ω_t remains a quadrature domain with simple time-dependence of the quadrature formula. Namely, choosing a test function u that is harmonic in all Ω_t, we have

$$\frac{d}{dt}\int_{\Omega_t} u(x)dV(x) = -Qu(x_0), \tag{21.11}$$

which completely determines the time-dependence of the quadrature formula and ensures that Ω_t remains a quadrature domain.

Richardson's Theorem states that the conservation law (21.11) holds more generally (not just for quadrature domains).

THEOREM 21.3 ([**160**]). *Suppose Ω_t solves Problem (21.1). Then (21.11) holds for any test function $u(x)$ harmonic in all Ω_t.*

In particular, take $n = 2$ and assume the sink is at the origin and that all Ω_t are simply connected. Using complex variable notation $z = x + iy$, it is common to apply Richardson's theorem to the cases $u(z) = z^k$ to obtain:

$$\frac{d}{dt}\int_{\Omega_t} z^k dA(z) = \begin{cases} -Q, & k = 0, \\ 0, & k = 1, 2, \ldots. \end{cases}$$

In other words, the so-called *harmonic moments* are all conserved except the zeroth moment, which corresponds to the area of the domain. This expresses infinitely many conservation laws. Moreover, the original problem of tracking the evolving domain can be reduced, in principle, to solving the inverse potential problem, the problem of recovering a domain from its harmonic moments, or equivalently from its logarithmic potential (which serves as a moment generating function in this context).

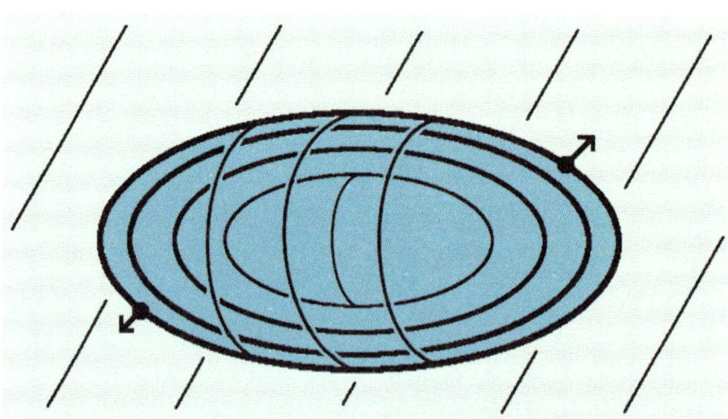

FIGURE 1. In the Laplacian growth problem, viscous fluid occupies the unbounded region and is removed through a sink at infinity. A homothetically increasing family of ellipsoids is an exact solution.

In the case referred to as *Laplacian growth*, where the viscous fluid domain is unbounded with a sink at infinity [**32**], an alternative formulation of Richardson's theorem states that the potential inside the cavity of the shell regions $\Omega_t \setminus \Omega_{s<t}$ is constant. It is then a consequence of Newton's Theorem (Corollary 16.2) that an increasing family of homothetic ellipsoids solves the Laplacian growth problem.

3. Exact solutions in the plane

Example 1 (Pascal's Limaçons): Consider the family of domains with boundaries given by the curves $\{\zeta : \zeta = az^2 + bz, |z| < 1\}$ with $a > 0$, $b > 2a$ real. The Schwarz function [**31**] is given by

$$S(\zeta) = -2ab/(a - \sqrt{a^2 + 4b\zeta}) + 4b^3/(a - \sqrt{a^2 + 4b\zeta})^2$$

which has a single-valued branch in the interior of the curve for appropriate parameter values a and b. The only singularities of the Schwarz function interior to the curve are a simple pole and a pole of order two at the origin. According to Remark 21.2, in order for a one-parameter family of domains to solve Problem (21.1), the singularities of $S(\zeta)$ must be time-independent except for one simple pole. Given an initial domain from this family we can choose a one-parameter family so that the residue at the origin decreases while the pole of order two does not change. This gives an exact solution to the Laplacian growth problem with suction taking place at the origin. The domain develops a cusp in finite time.

EXERCISE 21.1. Let Ω_0 be the Limaçon with $b = 10$ and $a = 1$. Write down a system of (nonlinear) equations for $a(t)$ and $b(t)$, in order for $\Omega(t)$ to solve Problem (21.1).

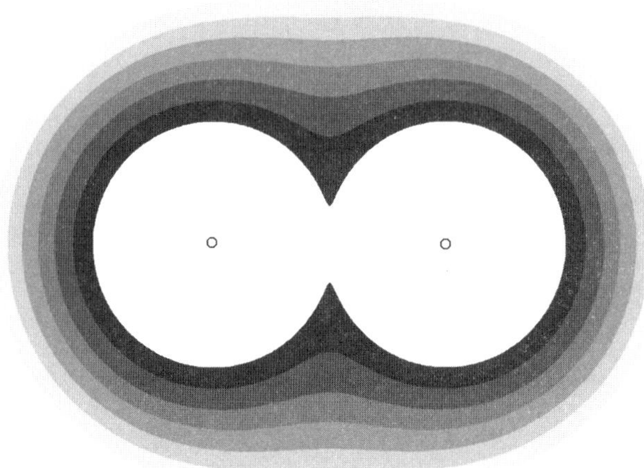

FIGURE 2. A family of Neumann ovals sharing the same pair of quadrature nodes.

Example 2 (C. Neumann ovals): Consider the family of domains with boundaries given by the solution set of the equation $(x^2+y^2)^2 = a^2(x^2+y^2)+4\varepsilon^2x^2$ (cf. Fig. 2). Then changing variables we have $(z\bar{z})^2 = a^2(z\bar{z}) + \varepsilon^2(z+\bar{z})^2$. Solving for $\bar{z}$ gives $S(z) = \frac{z(a^2+2\varepsilon^2)+z\sqrt{4a^4+4a^2\varepsilon^2+4\varepsilon^2z^2}}{2(z^2-\varepsilon^2)}$. We leave it as an exercise to verify that this is the correct choice of the branch of the square root. If ε is fixed, and $a(t)$ decreases such that the residues at $\pm\varepsilon$ decrease linearly, then the domain satisfies the multiple-sink version of (21.1) with sinks at $\pm\varepsilon$.

4. Algebraicity of planar quadrature domains

In the plane, under no regularity assumptions on $\partial\Omega$, it was shown by D. Aharanov and H. S. Shapiro [**5**] that quadrature domains are always algebraic, meaning the boundary $\partial\Omega$ is a subset of a real algebraic curve. B. Gustafsson (1983) showed that they have further nice properties [**67**] which we summarize in the following.

THEOREM 21.4 ([**67**]). *If $\Omega \subset \mathbb{R}^2$ is a quadrature domain, then Ω is algebraic, and, for some polynomial $P(x, y)$, the boundary $\partial\Omega$ consists of* **all** *of the points of $\{P(x, y) = 0\}$ except finitely many. Moreover, the leading order (homogeneous) term in P is a constant times $(x^2 + y^2)^M$, where M is determined by the order of the quadrature formula. Further, there are some explicit relations between the coefficients of P and the coefficients $a_{m,k}$ in the quadrature identity.*

The algebraicity of quadrature domains allows construction of exact solutions to the Laplacian Growth problem without resorting to conformal mapping. This motivated H.S. Shapiro to ask whether quadrature domains are always algebraic in $\mathbb{R}^n$, with $n \geq 3$. This question has a negative answer; A. Eremenko and the second author constructed four-dimensional examples with transcendental boundary. We present the construction in the next section.

5. Higher-dimensional quadrature domains need not be algebraic

Let us consider the case when the quadrature formula (21.9) is supported at a single point.

$$\int_\Omega u dV = \sum_{|\alpha|=0}^{M} a_\alpha \partial^\alpha u(0), \tag{21.12}$$

where $\alpha = (\alpha_1, \alpha_2, .., \alpha_n)$ again is a multi-index.

Under the condition that the leading coefficient $a_{(0,0,..,0)} > 0$ and the other coefficients a_α for $|\alpha| > 0$ are small in comparison, the existence and uniqueness of domains admitting a quadrature formula of the type (21.12) follows from a result on the solvability of the inverse potential problem [**179**] (cf. [**91**]).

Let us further specialize to the case when all derivatives in the quadrature formula (21.12) are with respect to the same variable, say x_1.

$$\int_\Omega u dV = \sum_{k=0}^{m} a_k \frac{\partial^k u}{\partial x_1^k}(0). \tag{21.13}$$

Prescribing the simplest nontrivial quadrature formula of the type (21.13), namely, where $m = 1$, $a_0 > 0$, and $a_1 > 0$, leads to non-algebraic examples [**47**].

THEOREM 21.5 ([**47**]). *In $\mathbb{R}^4$, there exist quadrature domains that are not algebraic, namely, admitting the quadrature formula (21.13) with $m = 1$, $a_0 > 0$, and $a_1 > 0$ (with a_1 small in comparison with a_0).*

We will work in terms of the Schwarz potential introduced in the previous section. Before beginning the proof we make an observation regarding axially symmetric potentials.

Recall the axially symmetric reduction of Laplace's equation:

Suppose that $u(x_1, x_2, ..., x_n)$ is harmonic in $\mathbb{R}^n$ and axially symmetric about the x_1-axis. Write $U(x, y) = u(x_1, x_2, ..., x_n)$, where $x = x_1$ and $y = \sqrt{x_2^2 + ... + x_n^2}$. Then,

$$\Delta U + \frac{n-2}{y} U_y = 0.$$

Indeed, $\Delta u = \operatorname{div}(U_x, U_y y_{x_2}, U_y y_{x_3}, ..., U_y y_{x_n})$

$$= U_{xx} + U_{yy}\left(\sum_{i=2}^{n} y_{x_i}^2\right) + U_y\left(\sum_{i=2}^{n} y_{x_i x_i}\right),$$

where easy calculations give $\displaystyle\sum_{i=2}^{n} y_{x_i}^2 = 1$ and $\displaystyle\sum_{i=2}^{n} y_{x_i x_i} = \frac{n-2}{y}$.

In the case $n = 4$, U satisfies the equation $\Delta U + \dfrac{2U_y}{y} = 0$ if and only if $yU(x, y)$ is a harmonic function of two variables (cf. Ch. 12). Indeed,

$$\Delta(yU) = y\Delta U + 2\nabla U \cdot \nabla y + U\Delta y = y\Delta U + 2U_y.$$

If Ω is a domain in $\mathbb{R}^4$ with axial symmetry, then the above considerations apply to the Schwarz potential, since the rotational symmetry of the Cauchy data (13.11) is passed to the solution w.

Thus, the Cauchy problem (13.11) defining the Schwarz potential $w(\mathbf{x})$ is reduced to the following two-dimensional Cauchy problem for the axially symmetric reduction $W(x, y)$ of $w(x_1, x_2, x_3, x_4)$:

$$\text{(21.14)} \qquad \begin{cases} \Delta W + \dfrac{2W_y}{y} = 0, \text{ near } \gamma, \\ W|_\gamma = \dfrac{1}{2}(x^2 + y^2), \\ \nabla W|_\gamma = (x, y), \end{cases}$$

where γ is the symmetric (about the x-axis) curve in the plane whose rotation generates the boundary of Ω, and (x, y) denotes the vector with components x and y.

According to the previous observation, we have

$$\text{(21.15)} \qquad \Delta\,(yW(x, y)) = 0,$$

which will be used below.

PROOF OF THEOREM 21.5. We will construct Ω by first describing a conformal map $f : \mathbb{D} \to D_\rho$ from the unit disk $\mathbb{D}$ to a domain D_ρ in the plane symmetric with respect to the real axis. Then we will take Ω to be the domain generated by rotation of D_ρ into $\mathbb{R}^4$ about the x-axis. i.e., $(x_1, x_2, x_3, x_4) \in \Omega$ if and only if $\left(x_1, \sqrt{x_2^2 + x_3^2 + x_4^2}\right) \in D_\rho$. Notice that the boundary $\Gamma = \partial\Omega$ is algebraic if and only if $\gamma = \partial D_\rho$ is algebraic, in which case $\gamma = \{\rho(x, y) = 0\}$, and

$\Gamma = \left\{\rho\left(x_1, \sqrt{x_2^2+x_3^2+x_4^2}\right) = 0\right\}$, where $\rho(x,y)$ is a polynomial that is even in the variable y.

The conformal map f and its relevant properties will be established in the proofs of the following claims.

Claim 1: There exists a function f with real coefficients, analytic and univalent in a neighborhood of $\overline{\mathbb{D}}$, with $f(0)=0$, so that the function

$$\left(f(z) - f\left(\frac{1}{z}\right)\right)^2 = g(z) \tag{21.16}$$

is rational, and its only pole in $\overline{\mathbb{D}}$ is a pole of exact order 3 at $z=0$.

Claim 2: The (analytically continued) function f has an infinitely-sheeted Riemann surface.

We defer the proofs of the claims in favor of first seeing how they are used to prove the theorem.

Since, by Claim 1, $D = f(\mathbb{D})$ is the image of $\mathbb{D}$ under a univalent function analytic in a neighborhood of $\overline{\mathbb{D}}$, the boundary of D_ρ, and therefore of Ω, is analytic and non-singular. Thus $\partial\Omega$ has a Schwarz potential $w(x_1,x_2,x_3,x_4)$, which by axial symmetry can be reduced to a function $W(x,y)$ of two variables (see the discussion above just before the proof). As stated in (21.15), $yW(x,y)$ is harmonic near ∂D_ρ.

Let $\zeta = x+iy$. Then the function

$$V(x,y) = yW(x,y) - Im\{\zeta^3\}/6,$$

is harmonic (subtracting the harmonic term $Im\{\zeta^3\}/6$ simplifies the following calculations without changing the singularities of V).

Next, consider the (complex-analytic) function that results from taking the ζ-derivative $V_\zeta = \frac{1}{2}(V_x - iV_y)$.

$$V_\zeta = \frac{-i}{2}W + yW_\zeta + \frac{i\zeta^2}{4}.$$

The Cauchy data for W, stated in (21.14), can be written as $W = \zeta\bar{\zeta}/2$ and $W_\zeta = \bar{\zeta}/2$, so that on the boundary $\gamma = \partial D_\rho$, the equation above gives

$$V_\zeta|_\gamma = -\frac{i\zeta\bar{\zeta}}{4} + \frac{\zeta-\bar{\zeta}}{4i}\cdot\bar{\zeta} + \frac{i\zeta^2}{4} = \frac{i}{4}\left(\zeta-\bar{\zeta}\right)^2,$$

where we have also used $y = \dfrac{\zeta-\bar{\zeta}}{2i}$.

Now we replace $\bar{\zeta}$ with $S(\zeta)$, the Schwarz function of γ. Then, both sides of the equation are analytic in a neighborhood of γ, and thus the following becomes an identity (not just valid on the boundary).

$$V_\zeta = \frac{i}{4}\left(\zeta - S(\zeta)\right)^2. \tag{21.17}$$

This relates the four-dimensional Schwarz potential to the two-dimensional Schwarz function $S(\zeta)$ which can be written in terms of the conformal map f

$$S(\zeta) = f^*\left(\frac{1}{f^{-1}(\zeta)}\right), \tag{21.18}$$

where f^* denotes the function obtained by conjugating the coefficients of f.

As follows from (21.18) when the coefficients of f are real, the Schwarz function of γ satisfies the functional equation

$$S(f(z)) = f\left(\frac{1}{z}\right).$$

Using this relationship, and substituting $\zeta = f(z)$ into (21.17), we get

$$V_\zeta(f(z)) = \frac{i}{4}\left(f(z) - f\left(\frac{1}{z}\right)\right)^2.$$

By Claim 1 the right hand side is $g(z)$, a function analytic in a neighborhood of $\overline{\mathbb{D}}$, except for a pole at $z = 0$ of order 3. So, $g(z) = \dfrac{Q(z)}{z^3}$, with $Q(z)$ analytic in a neighborhood of $\overline{\mathbb{D}}$ and $Q(0) \neq 0$.

Since $f(0) = 0$ and f is univalent, we have $f^{-1}(\zeta) = \zeta h(\zeta)$, for some $h(\zeta)$ analytic and non-vanishing in D_ρ. Thus,

$$V_\zeta(\zeta) = g(f^{-1}(\zeta)) = \frac{1}{\zeta^3}\frac{Q(f^{-1}(\zeta))}{h(\zeta)^3},$$

which is analytic in D_ρ except for a pole of order 3 at $\zeta = 0$. Moreover, expanding about $\zeta = 0$ we have, for some $p(\zeta)$ analytic in D_ρ and $A > 0$ and $B > 0$,

$$V_\zeta(\zeta) = -A\frac{i}{\zeta^2} + B\frac{i}{\zeta^3} + p(\zeta), \tag{21.19}$$

where the residue of V_ζ is zero, as can be seen by integrating (21.17) over ∂D_ρ, and noting that on this contour, with respect to reflection over the x-axis, the integrand is even and the differential $d\zeta$ is odd.

In order to recover $W(x, y)$ we need to take the real part of a primitive of (21.19) and divide by y (cf. [**96**, Lemma 2.3]). A primitive of (21.19) is

$$A\frac{i}{\zeta} - \frac{B}{2}\frac{i}{\zeta^2} + P(\zeta), \tag{21.20}$$

with $P(\zeta)$ analytic in D_ρ.

Notice that

$$Re\left\{\frac{i}{\zeta}\right\} = \frac{y}{x^2 + y^2},$$

and

$$Re\left\{\frac{i}{\zeta^2}\right\} = -y\frac{\partial}{\partial x}\frac{1}{x^2 + y^2}.$$

This implies that (cf. [**96**, p. 183])

$$w(x_1, x_2, x_3, x_4) = w(x) = A \cdot ||x||^{-2} + \frac{B}{2} \cdot \partial_{x_1}(||x||^{-2}) + H(x),$$

where $H(x)$ is harmonic in Ω.

Since, up to a constant, $||x||^{-2}$ is the fundamental solution $E_4(x, \mathbf{0})$ evaluated at $\mathbf{y} = \mathbf{0}$, Theorem 17.2 implies that Ω is a quadrature domain with quadrature formula of the form

$$\int_\Omega u\,dV = a_0 u(\mathbf{0}) + a_1\frac{\partial u}{\partial x_1}(\mathbf{0}).$$

Next we apply Claim 2 in order to show that the boundary of Ω is not algebraic.

Note: It does not follow immediately from Claim 2 that the image of $\partial\mathbb{D}$ is non-algebraic, since it is possible for an algebraic domain to be conformally mapped from the unit disk by a transcendental function. Indeed, any non-circular ellipse is such an example.

Suppose, to the contrary, that $\gamma = \partial D_\rho$ is the zero set of a polynomial $\rho(x, y)$. Then the Schwarz function $S(\zeta)$ is an algebraic function, since it can be obtained by solving for $\bar{\zeta}$ in the equation

$$\rho\left(\frac{\zeta+\bar{\zeta}}{2}, \frac{\zeta-\bar{\zeta}}{2i}\right) = 0.$$

Substitute $z = f^{-1}(\zeta)$ into (21.16) and use the formula (21.18)

$$S(\zeta) = f\left(\frac{1}{f^{-1}(\zeta)}\right),$$

where again, the conjugation is missing because the coefficients of f are real. We obtain

$$g(f^{-1}(\zeta)) = (\zeta - S(\zeta))^2.$$

Since $g(z)$ is rational, and $S(\zeta)$ is algebraic, this implies

$$f^{-1}(\zeta) = g^{-1}\left((\zeta - S(\zeta))^2\right)$$

is also an algebraic function. This contradicts the fact that $f^{-1}(\zeta)$ has an analytic continuation to an infinitely sheeted Riemann surface. We conclude that $S(\zeta)$ is not an algebraic function, and so the boundary of Ω is not contained in the zero set of a polynomial.

□

It remains to prove the claims.

PROOF OF CLAIM 1. With $0 < a < 1$ a real parameter, let

$$h(z) = \frac{z^2-1}{z}\sqrt{\frac{z+a}{z}(1+az)}.$$

It is easy to see that the square root above has two single valued analytic branches on $|z| = 1$. We choose that branch which is > 0 at $z = 1$.

We observe that

$$h(1/z) = -h(z). \tag{21.21}$$

Since $h(z)$ is analytic in a neighborhood of $\partial\mathbb{D}$, say $\{1 - \varepsilon < |z| < 1 + \varepsilon\}$ it has a Laurent expansion

$$h(z) = \sum_{j=-\infty}^{\infty} c_j z^j.$$

Let $f(z) = \sum_{j=1}^{\infty} c_j z^j$. By (21.21), $c_j = -c_{-j}$.

Therefore,

$$h(z) = f(z) - f\left(\frac{1}{z}\right), \tag{21.22}$$

where $f(z)$ is analytic in a neighborhood of $\overline{\mathbb{D}}$, and $f(1/z)$ is analytic in a neighborhood of $\mathbb{C} \setminus \mathbb{D}$ with $f(1/z) \to 0$ as $z \to \infty$.

Observe that $f(z)$ satisfies (21.16) if we take

$$g(z) = h(z)^2 = \frac{(z^2-1)^2(z+a)(1+az)}{z^3}.$$

We remark that, up to multiplication by a constant (resulting in a dilation of the domain) and changing the sign of the parameter a (resulting in a reflection of the domain), this is the only choice of g for which all the conditions in the claim can be satisfied.

Since f contains exactly the terms with positive powers in the Laurent expansion of h, it follows that for $w \in \mathbb{D}$, $f(w)$ can be expressed as the contour integral

$$f(w) = \frac{1}{2\pi i} \int_{|z|=1} \frac{h(z)}{z-w} dz.$$

So we have

$$f(w) = \frac{1}{2\pi i} \int_{|z|=1} \frac{(z^2-1)\sqrt{(1+\frac{a}{z})(1+az)}}{z(z-w)} dz. \tag{21.23}$$

Next, we show that $f(w)$ is univalent at least for sufficiently small values of the parameter $0 < a < 1$.

Use (21.23) to write the derivative $f'(w)$:

$$f'(w) = \frac{-1}{2\pi i} \int_{|z|=1+\varepsilon} \frac{(z^2-1)\sqrt{(1+\frac{a}{z})(1+az)}}{z(z-w)^2} dz,$$

where we have also deformed the contour slightly with $\varepsilon > 0$. As $a \to 0$, the integrand converges (uniformly for $|z| = 1+\varepsilon$) to $\frac{(z^2-1)}{z(z-w)^2}$. Thus, $f'(w)$ converges uniformly in $\overline{\mathbb{D}}$ to

$$\frac{1}{2\pi i} \int_{|z|=1+\varepsilon} \left(\frac{z}{(z-w)^2} - \frac{1}{z(z-w)^2} \right) dz = 1,$$

using residues.

This guarantees univalence of $f(w)$ throughout $\overline{\mathbb{D}}$ for all sufficiently small $a > 0$. Indeed, for any two points w_1 and w_2 in $\overline{\mathbb{D}}$,

$$f(w_2) - f(w_1) = \int_{w_1}^{w_2} f'(w) dw,$$

and $f'(w)$ is uniformly close to 1 for $a > 0$ sufficiently small, making the left-hand-side non-zero for $w_1 \neq w_2$.

□

Proof of Claim 2. We investigate the global analytic function whose principal branch is given by the integral (21.23).

We will deform the contour in (21.23), but first we manipulate the integrand so that the new integral will converge. Using

$$\frac{(z^2-1)}{z(z-w)} = 1 + \frac{w}{z} + \frac{w^2-1}{z(z-w)},$$

we have

$$f(w) = A_0 + A_1 w + (w^2-1)F(w),$$

where A_0 and A_1 are integrals that do not depend on w, and

$$F(w) = \frac{1}{2\pi i} \int_{|z|=1} \frac{\sqrt{(1+\frac{a}{z})(1+az)}}{z(z-w)} dz. \tag{21.24}$$

Now replace the contour with a keyhole contour consisting of a circle of large radius R traced counterclockwise but where a small arc of the circle near angle π is replaced by a detour, namely, a path traveling around the segment $(-\infty, -1/a]$. The contribution along the large circle is $O(R^{-1})$, and so letting $R \to \infty$ we can replace the contour by one that "just surrounds" the segment $(-\infty, -1/a]$. The integrand changes sign as the path switches from the top side (which is traced from left to right) to the bottom side of the segment (which is traced from right to left). Thus, we obtain twice the integration along a segment:

$$F(w) = \frac{1}{\pi i}\int_{-\infty}^{-1/a} \frac{\sqrt{(1+\frac{a}{z})(1+az)}}{z(z-w)}dz = \frac{1}{\pi i}\int_{-\infty}^{-1/a} \frac{\sqrt{G(z)}}{(z-w)}dz, \tag{21.25}$$

where $G(z) = (1+\frac{a}{z})(\frac{1}{z}+a)\frac{1}{z}$. We use (21.25) in order to show that the analytic continuation of $F(w)$, and hence $f(w)$, has an infinitely-sheeted Riemann surface.

The monodromy for $F(w)$ can be obtained as a nice exercise in applying the Sokhotski-Plemelj relation [**82**, Ch. 14]. Accordingly, the jump across the segment $(-\infty, -1/a]$ of such an integral as appears in (21.25) is equal to $2\pi i$ times the numerator of the integrand, i.e., let w approach a point x on the segment $(-\infty, -1/a]$ from above and below, respectively, then

$$F(x_+) - F(x_-) = -2\sqrt{G(x)}. \tag{21.26}$$

In order to perform analytic continuation, choose a base point w_0, say in the upper half-plane, and let $F_0(w)$ denote the principal branch (which is expressed by the integral itself). Then, let γ_1 be a loop based at w_0 that winds once around $-1/a$. According to (21.26), performing analytic continuation along γ_1, we obtain a new branch $F_1(w) = F_0(w) + 2\sqrt{G(w)}$. If γ_1 is traced again, then we return to the original branch since $2\sqrt{G(w)}$ switches sign and cancels the contribution from F_0. So $w = -1/a$ only has ramification index 2.

Consider a second loop γ_2 based at w_0 that winds around the origin, crossing the segment $[0, a]$. Traveling along this loop does nothing to the initial branch F_0, but if we first continue along γ_1, then the branch F_1 will be affected when we continue along γ_2. Alternate between these two loops and perform analytic continuation along $\gamma_1 \cdot \gamma_2 \cdot \gamma_1 \cdots \gamma_1$, where γ_1 is traced $k+1$ times. Let $F_i(w)$ denote the new branch obtained at each step. Then we have

$$\begin{aligned} F_1(w) &= F_0(w) + 2\sqrt{G(w)}, \\ F_2(w) &= F_0(w) - 2\sqrt{G(w)}, \\ F_3(w) &= F_0(w) + 2\sqrt{G(w)} + 2\sqrt{G(w)}, \\ F_4(w) &= F_0(w) - 4\sqrt{G(w)}, \\ F_5(w) &= F_0(w) + 2\sqrt{G(w)} + 4\sqrt{G(w)}, \ldots, \end{aligned}$$

and, in general,

$$F_{2k+1}(w) = F_0(w) + 2(k+1)\sqrt{G(w)}.$$

Let us elaborate on the first couple of steps in the process described in the equations displayed above. Going from $F_1(w)$ to $F_2(w)$, the sign of $2\sqrt{G(w)}$ changed since γ_2 loops around the origin and G has a pole of odd order 3 at the origin. Then, going from $F_2(w)$ to $F_3(w)$, F_0 gets an increment of $2\sqrt{G(w)}$ and the sign of the original term $2\sqrt{G(w)}$ changes again since G has a zero of order one at $-1/a$. Thus, the Riemann surface for $F(w)$ is infinitely-sheeted. $\square$

Notes

The question on algebracity of higher-dimensional quadrature domains was raised by H. S. Shapiro [**176**] (p. 40) in 1992 and posed again by B. Gustafsson in the open problems section of [**39**] (p. xv). The answer presented here was given in [**47**]. Construction of a non-trivial quadrature domain in three dimensions remains an open problem. The inverse potential problem is underdetermined; there are known examples of non-uniqueness—distinct domains having the same harmonic moments. (It is an open problem posed in [**5**] to determine whether uniqueness holds for the inverse potential prob- lem within the restricted class of quadrature domains.) However, in the Laplacian growth problem the final domain is uniquely determined by the initial domain along with the description of the evolution of harmonic moments. See [**195**] for details. The axially symmetric reduction of Laplace's equation was used by L. Karp [**96**] to construct quadrature domains in $\mathbb{R}^4$. For a discussion of more general growth processes governed by elliptic operators with variable coefficients, see [**108**] and the next chapter which surveys other types of quadrature domains and further applications in fluid dynamics.

CHAPTER 22

Other Varieties of Quadrature Domains

The definition of a quadrature domain introduced in the previous chapter may be modified in several interesting directions. For instance, in place of a sum of point evaluations, one could consider quadrature identities involving more general distributions. Instead of integrating with respect to uniform volume measure over the domain, one could use weighted measure, or also restrict the support of the measure, say surface measure over the boundary. Instead of harmonic functions, an alternative class of test functions could be used.

In the sections below we will present several examples illustrating various alternative notions of quadrature domains. In the Notes section at the end, we conclude by mentioning several applications where quadrature domains appear as exact solutions to problems in fluid dynamics.

1. Ellipsoids as quadrature domains in the wide sense

Let us refer to a domain Ω as a *quadrature domain in the wide sense* if there is a distribution T supported on a compact subset of Ω so that for all u, say, harmonic in Ω and continuous in $\overline{\Omega}$

$$\int_\Omega u dV = \langle u, T\rangle. \tag{22.1}$$

As a corollary of MacLaurin's Theorem (using exactly the same proof as in Corollary 15.2), we notice that a heterogeneous ellipsoid

$$\Omega := \left\{ x \in \mathbb{R}^N : \sum_{j=1}^N \frac{x_j^2}{a_j^2} - 1 = 0, a_1 > a_2 > \cdots \right\},$$

is a quadrature domain in the wide sense, where the distribution T is a continuous function times the $(N-1)$-dimensional Lebesgue measure supported on the focal ellipsoid E.

$$\int_\Omega u dV = \int_E u d\mu, \tag{22.2}$$

where

$$d\mu(x) = 2\left(\prod_{j=1}^N a_j\right)\left(\prod_{j=1}^{N-1}(a_j^2 - a_N^2)\right)^{-1/2}\left(1 - \sum_{j=1}^{N-1}\frac{x_j^2}{a_j^2 - a_N^2}\right)^{1/2} dA(x).$$

2. Null quadrature domains

One may also allow unbounded domains (which show up naturally in the Laplacian growth problem). Of particular interest are the so-called *null quadrature domains*, unbounded quadrature domains for which the quadrature formula vanishes. That is, for all harmonic test functions $u \in L^1(\Omega)$, we have

$$\int_\Omega u dV = 0.$$

In two dimensions, M. Sakai [**164**] classified null quadrature domains, showing that the class consists entirely of exteriors of ellipses and other regions that can be obtained as degenerations of such (half-planes, the non-convex region bounded by a parabola, and the complement of a line segment). Similar results have been established in higher dimensions under additional assumptions [**54**].

The following exercise addresses a question that was asked in the preface.

EXERCISE 22.1. (a) Prove that the non-convex region $\Omega := \{(x, y) \in \mathbb{R}^2 : x < y^2\}$ bounded by a parabola is a null quadrature domain for analytic functions. Hint: Since Ω is simply connected, all harmonic functions are real parts of analytic functions. Thus, it is more convenient to take analytic test functions. As a basis for such test functions one may take functions of the form $\frac{1}{(z-z_0)^k}$ with z_0 in $\mathbb{C} \setminus \overline{\Omega}$ and $k \geq 3$ an integer [**20**]. Show that the area integral over Ω of such a function vanishes by using the Cauchy-Green Theorem (cf. Remark 13.2) and an explicit computation of the Schwarz function $S(z)$ of the parabola.

(b) Prove that the non-convex region $U := \{(x, y) \in \mathbb{R}^2 : x^2 - y^2 < 1\}$, bounded by a hyperbola, is not a null quadrature domain [**164**]. Hint: As a test function choose $f(z) = (2 - z^2)^{-3/2}$, which has a single-valued branch in U that is area integrable. Use the Cauchy-Green Theorem and the Schwarz function $S(z)$ of the hyperbola to rewrite the integral $\int_U f(z)dA(z)$ as a contour integral. Writing this as two separate integrals (one over each branch of the hyperbola), the separate integrals can each be computed, after simplification, using residues (since each integrand simplifies to a rational function). Note that the two components of the hyperbola give rise to two different branches of the Schwarz function $S(z) = \pm\sqrt{2 - z^2}$.

3. Arclength quadrature domains

By considering integration over the boundary of the domain, one arrives at the notion of a surface quadrature domain, or restricting attention to the planar case, we say that Ω is an *arclength quadrature domain* for analytic functions if the identity

$$\int_{\partial\Omega} f ds = \sum_{k=1}^{n} c_k f(z_k)$$

holds for all f in some test class, say for all f in the Smirnov class $E^1(\Omega)$ (see [**36**]).

One may also consider *arclength null-quadrature domains*, where the domain Ω is unbounded and the right hand side of the quadrature identity vanishes. The classification of such domains remains an open problem that has proven to be more intricate than the classification problem for area null quadrature domains. Partial

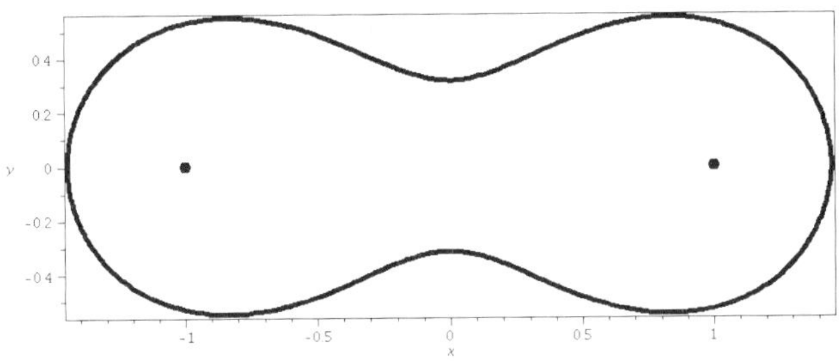

FIGURE 1. An oval of Cassini given by $\{z \in \mathbb{C} : |z^2 - 1| = 1.1\}$.

results and constructions have used transcendental [**76**], [**13**] and non-elementary [**48**] conformal mapping, conformal mappings whose derivatives are purely singular inner functions [**107**], half-order and meromorphic differentials on Riemann surfaces [**68**], [**48**], and a non-trivial connection to minimal surfaces [**191**], [**192**].

In fact, most of the above studies assume the existence of a so-called "roof function", a positive, harmonic function with constant Dirichlet data (possibly a different constant on each boundary component) and constant Neumann data (the same constant on each boundary component). This condition is sufficient for a domain to be a null-quadrature domain, but whether it is necessary is an open question, that is, does every arclength null-quadrature domain admit a roof function?

Moreover, the problem of characterizing domains for which the quadrature distribution is not compactly supported, e.g., the quadrature distribution is simply arclength on the boundary, i.e.,

$$\int_\Omega f dA = \frac{\text{Area}(\Omega)}{\text{Perim}(\Omega)} \int_{\partial\Omega} f ds,$$

for all analytic in Ω functions f, has interesting connections to many other problems in mathematical physics united under the umbrella of "free boundary problems". Normally, these problems admit very few solutions but are excruciatingly difficult, and each one requires special methods of attack. Very few of those are solved - cf. [**1**], for the final resolution of the above problem, some recent advances and many relevant references.

4. Lemniscates as quadrature domains for equilibrium measure

The ovals of Cassini $\{z \in \mathbb{C} : |z^2 - 1| < T\}$ (cf. Fig. 1) are not quadrature domains in the classical sense. However the exercises below show that they may be viewed as quadrature domains for equilibrium measure as well as quadrature domains in the wide sense and quadrature domains for a certain weighted area measure.

First we will discuss more generally lemniscate domains. Consider the domain $\Omega \subset \mathbb{C}$ defined by the sublevel set of the modulus of a monic polynomial $p(z) = \prod_{k=1}^n (z - z_k)$, namely,

$$\Omega := \{z \in \mathbb{C} : |p(z)| < T\}.$$

This set is connected and thus defines a domain if T exceeds the modulus of the largest in modulus critical value of $p(z)$.

We can interpret Ω as a quadrature domain with respect to so-called *equilibrium measure*. The equilibrium measure of Ω, which in general is defined in terms of a certain energy minimization problem from logarithmic potential theory [**163**], reduces in this case to

$$d\omega = \frac{1}{n}\partial_n \log|p(z)|ds,$$

where ∂_n denotes the derivative in the outward normal direction, and ds denotes arclength measure on the boundary $\partial\Omega$ of Ω. Integrating with respect to $d\omega$, we have the following quadrature identity, where the test function u is, say, any function harmonic in a neighborhood of $\overline{\Omega}$.

$$\int_{\partial\Omega} u\,d\omega = \frac{1}{n}\sum_{k=1}^{n} u(z_k). \tag{22.3}$$

EXERCISE 22.2. Prove the quadrature identity (22.3). Hint: Use the fact that $v(z) = \frac{1}{n}(\log|p(z)| - \log T)$ vanishes on $\partial\Omega$ to rewrite the left hand side of (22.3) in a form that is set up for using Green's second identity. Note that, since v has singularities in Ω, this requires either using a distributional form of Green's second identity, or (for a more elementary approach) introducing small circular contours centered at each z_k.

EXERCISE 22.3. Fix $T > 1$, and consider the domain $\Omega = \{z \in \mathbb{C} : |z^2 - 1| < T\}$ which is bounded by an oval of Cassini.

(a) Prove that Ω can be viewed as yet another type of quadrature domain involving a weighted measure, namely, for area measure weighted by the continuous density function $|z|^2$. For every harmonic test function u,

$$\int_{\Omega} u(z)|z|^2 dA(z) = u(-1) + u(1).$$

Hint: Use the complex form of Green's Theorem, and find an explicit formula for the Schwarz function of $\partial\Omega$.

(b) Cassini's oval is also a quadrature domain in the wide sense (for area measure) where the distribution is supported on the segment $[-1, 1]$ of the real axis. Using the Schwarz function of $\partial\Omega$, determine a real-valued function $\gamma(x)$ so that for each harmonic test function u

$$\int_{\Omega} u(z)dA(z) = \int_{-1}^{1} u(x)\gamma(x)dx.$$

Another interesting free boundary problem stems from a general "philosophical" query of finding the domains such that naturally defined measures associated with those domains have some sort of dependency. More precisely, there are three natural measures "living" on the boundary of any smoothly bounded simply connected bounded domain: the harmonic measure (cf. [**153**], [**163**], [**6**]), arclength, and the angle subtended by the arc of the boundary viewed from a fixed point inside the domain, say the origin. A natural question asked by H.S. Shapiro and the first author in the '80s is for what domains is one of these three measures a linear combination of the other two? The question remains open in its entirety. H. S. and DK proved (unpublished) that if any two of the three measures are scalar multiples of one another, the domain must be a disk (or, a ball in higher dimensions). This is a satisfying exercise in elementary function theory and is left to the reader. Yet, this is all that is known. One, of course, can modify the question and ask, e.g., for which domains the linear combination of the three boundary measures serves as the quadrature with respect to area for, say analytic or harmonic test functions. This all remains a totally untraveled territory.

5. Quadrature domains for other classes of test functions

Instead of analytic or harmonic functions, let us consider a test class of functions u satisfying a divergence-form elliptic PDE

$$\operatorname{div}(\lambda \nabla u) = 0.$$

Thus, we refer to Ω as an *elliptic quadrature domain* if

$$\int_\Omega u(z)dA(z) = \sum_{k=1}^{n} a_k u(z_k) \tag{22.4}$$

is satisfied for every function u that solves $\operatorname{div}(\lambda \nabla u) = 0$ (say in a neighborhood of $\overline{\Omega}$).

As a simple example (yet one that already leads to a domain with transcendental boundary), we will take $\lambda = x^2$ and construct a domain that admits the simplest possible quadrature identity

$$\int_\Omega u(z)dA(z) = \alpha u(1), \tag{22.5}$$

with one simple node positioned at the point $z = 1$ (note the resemblance with the mean value property).

We describe the domain by specifying its conformal mapping $f : \mathbb{D} \to \Omega$ from the unit disk to Ω. We will choose f to satisfy the functional equation

$$f(w) + f\left(\tfrac{1}{w}\right) = 2\gamma \exp\left\{\beta\left(w + \frac{1}{w}\right)\right\}, \tag{22.6}$$

where β is a parameter whose relation to α will be explained below, and $\gamma = \gamma(\beta)$ will be chosen so that $f(0) = 1$.

Since f is analytic in the unit disk, its Taylor series can be extracted from the functional equation (22.6) by taking the Laurent expansion of the right hand side, which is valid throughout $\mathbb{C} \setminus \{0\}$, and then simply taking $f(w)$ to be the part of this expansion with positive powers (including half of the constant term). Thus, the Taylor expansion of f is given by

$$f(w) = \gamma\left(\sum_{k=0}^{\infty} \frac{\beta^{2k}}{(k!)^2} + \sum_{m=1}^{\infty}\sum_{k=0}^{\infty} \frac{2\beta^{m+2k} w^m}{(m+k)!k!}\right),$$

where $\gamma = \left(\sum_{k=0}^{\infty} \frac{\beta^{2k}}{(k!)^2}\right)^{-1}$ is chosen in order to have $f(0) = 1$.

Note from the Taylor expansion that f is entire (the first series is is nothing but $J_o(2i\beta)$, where J_0 is the zeroth Bessel function). The functional equation (22.6) implies $\log[f(w) + f(\frac{1}{w})] = \beta\left(w + \frac{1}{w}\right) + \log(2\gamma)$.

In order to verify the quadrature identity (22.4), let u be an arbitrary test function satisfying $\operatorname{div}(\lambda \nabla u) = 0$ in a neighborhood of $\overline{\Omega}$. Set $v(z) = xu(z)$, and observe that v is harmonic, since $\operatorname{div}(\lambda \nabla u) = x\Delta(xu)$. Since Ω is simply connected, $v(z) = Re V(z)$ where $V(z)$ is analytic. Then we have:

$$\begin{aligned}\int_\Omega u(z)dA(z) &= \int_\Omega \frac{1}{x} v(z)dA(z) \\ &= Re \int_\Omega \frac{1}{x} V(z)dA(z) \\ &= Re \frac{1}{i} \int_{\partial\Omega} \log(z+\bar{z})\, V(z)dz,\end{aligned}$$

where we have used Green's Theorem in the last equation.

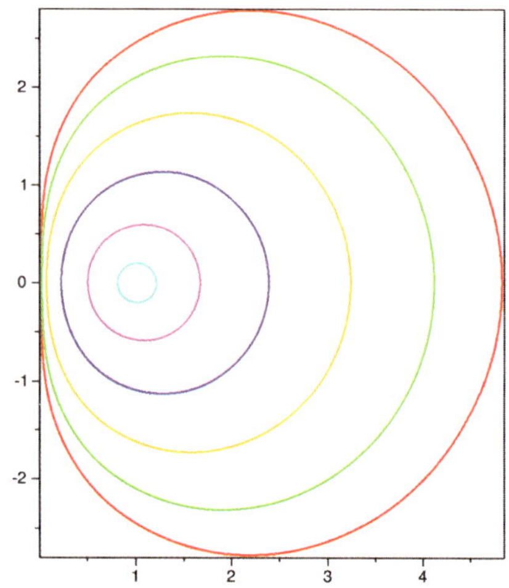

FIGURE 2. Elliptic quadrature domains for $\lambda = x^2$ with one simple quadrature node positioned at $z = 1$. In these examples, $\beta = 0.1, 0.3, 0.6, 1, 1.5, 2$.

After making a change of variables $z = f(w), dz = f'(w)dw$, we obtain

$$\begin{aligned}Re \frac{1}{i} \int_{\partial\Omega} \log(z+\bar{z})\, V(z)dz &= Re \frac{1}{i} \int_{\partial\mathbb{D}} \log\left(f(w) + f\left(\tfrac{1}{w}\right)\right) V(f(w))f'(w)dw \\ &= Re \frac{1}{\pi i} \int_{\partial\mathbb{D}} \pi \left[\beta\left(w + \tfrac{1}{w}\right) + \log(2\gamma)\right] V(f(w))f'(w)dw \\ &= Re\left\{2\pi\beta V(f(0))f'(0)\right\} \\ &= 2\pi\beta f'(0)v(1) \\ &= 2\pi\beta f'(0)u(1),\end{aligned}$$

where we have used the Cauchy integral theorem and the Cauchy integral formula to evaluate the integral. This establishes (22.5) with the value of the coefficient α given by

$$\alpha = 2\pi\beta f'(0) = 4\pi\gamma\beta^2 \sum_{k=0}^{\infty} \frac{\beta^{2k}}{(k+1)!k!}.$$

We leave showing that f is univalent in the unit disk as an exercise for the reader; under the additional assumption that $\beta > 0$ is sufficiently small, one may use the technique in the proof of Theorem 21.5 from the previous chapter. Namely, it suffices to estimate $f'(w)$ and show that it is uniformly close to a non-zero constant. On the other hand, the exercise is more challenging when β is not assumed small.

Some examples of these elliptic quadrature domains are plotted in Fig. 2. Notice that they do not cross the characteristic line $\{x = 0\}$.

For more details on elliptic quadrature domains and so-called elliptic growth processes, the reader may confer [**108**] and the references therein.

Notes

Null quadrature domains for area measure arise naturally in the Laplacian growth problem, where they appear as the only initial domains that are completely emptied by a sink at infinity [**32**].

Elliptic quadrature domains, in combination with the notion of integrating with respect to a weight function, appear in generalizations of the Hele-Shaw problem that account for a porous medium that has non-constant coefficients of "filtration" and "porosity" [**49**], [**134**], [**108**], [**136**].

Both area and arclength quadrature domains have appeared as exact solutions to problems in vortex dynamics. Namely, in certain steadily rotating vortex arrays area quadrature domains appear as regions of constant vorticity that represent vortex cores [**27**]. Arclength quadrature domains, and especially arclength null quadrature domains, appear in a separate setting where vortex cores are modelled by regions of constant pressure surrounded by a free streamline. The streamline is required to be at equilibrium given the pressure determined (through Bernoulli's law) by the fluid velocity along the streamline, [**132**], [**48**].

Arclength quadrature domains appear in several related problems involving a free streamline, for instance, where the equilibrium condition is that the force determined by Bernoulli's law is balanced by interfacial surface tension [**46**]. In certain cases, the solutions turn out to be *double quadrature domains*, that is, quadrature domains for both area and arclength measure [**30**].

Quadrature domains also appear in problems involving slow viscous flows, such as in viscous sintering, where the fluid stream function solves the (fourth-order) bilaplace equation [**29**].

For further applications of quadrature domains in fluid dynamics, see the survey [**28**].

It was shown in [**109**] that lemniscates are not preserved by Laplacian growth. On the other hand, certain lemniscates do survive certain "elliptic growth" processes as was shown in [**138**].

Although (classical sense) quadrature domains are special, they are not too rare. It is shown in [**16**] that they are ubiquitous in the space of finitely-connected smoothly bounded domains. A related density result is proved for the above-mentioned double quadrature domains in [**19**].

Quadrature domains appear in yet another direction of operator theory, namely, in the context of subnormal operators with finite-rank self-commutator [**141**], (see also the recent text [**69**]).

Bibliography

[1] A. Abanov, C. Beneteau, D. Khavinson, and R. Teodorescu, *A free boundary problem associated with the isoperimetric inequality, preprint (2016), arxiv:1601.03885*, 2016, to appear in J. d'Analyse Math.

[2] D. Aberra, *On a generalized reflection law for functions satisfying the Helmholtz equation*, Electron. J. Differential Equations (1999), No. 20, 10. MR1692561

[3] D. Aberra, *A note on the Schwarz reflection principle for polyharmonic functions*, Analysis (Munich) **31** (2011), no. 3, 205–210. MR2822305

[4] D. Aberra and T. Savina, *The Schwarz reflection principle for polyharmonic functions in* $\mathbf{R}^2$, Complex Variables Theory Appl. **41** (2000), no. 1, 27–44. MR1758596

[5] D. Aharonov and H. S. Shapiro, *Domains on which analytic functions satisfy quadrature identities*, J. Analyse Math. **30** (1976), 39–73. MR0447589

[6] L. V. Ahlfors, *Complex analysis*, 3rd ed., McGraw-Hill Book Co., New York, 1978. An introduction to the theory of analytic functions of one complex variable; International Series in Pure and Applied Mathematics. MR510197

[7] J. Armel and P. Ebenfelt, *Extension of solutions to holomorphic partial differential equations*, Comm. Partial Differential Equations **39** (2014), no. 9, 1770–1779. MR3246043

[8] D. H. Armitage, *The Dirichlet problem when the boundary function is entire*, J. Math. Anal. Appl. **291** (2004), no. 2, 565–577. MR2039070

[9] V. I. Arnold, *Magnetic analogs of the theorems of newton and ivory*, Uspekhi Mat. Nauk, 38 (1984), pp. 253–254.

[10] N. Aronszajn, T. M. Creese, and L. J. Lipkin, *Polyharmonic functions*, Oxford Mathematical Monographs, The Clarendon Press, Oxford University Press, New York, 1983. Notes taken by Eberhard Gerlach; Oxford Science Publications. MR745128

[11] L. Asgeirsson, *Über Mittelwertgleichungen, die mehreren partiellen Differentialgleichungen 2. Ordnung zugeordnet sind* (German), Studies and Essays Presented to R. Courant on his 60th Birthday, January 8, 1948, Interscience Publishers, Inc., New York, 1948, pp. 7–20. MR0022986

[12] V. Avanissian, *Cellule d'harmonicité et prolongement analytique complexe* (French), Travaux en Cours. [Works in Progress], Hermann, Paris, 1985. MR851008

[13] G. R. Baker, P. G. Saffman, and J. S. Sheffield, *Structure of a linear array of hollow vortices of finite cross-section*, Journal of Fluid Mechanics, 74 (1976), pp. 469–476.

[14] H. Behnke and P. Thullen, *Theorie der Funktionen mehrerer komplexer Veränderlichen* (German), Ergebnisse der Mathematik und ihrer Grenzgebiete, Band 51. Zweite, erweiterte Auflage. Herausgegeben von R. Remmert. Unter Mitarbeit von W. Barth, O. Forster, H. Holmann, W. Kaup, H. Kerner, H.-J. Reiffen, G. Scheja und K. Spallek, Springer-Verlag, Berlin-New York, 1970. MR0271391

[15] B. P. Belinskiy and T. V. Savina, *The Schwarz reflection principle for harmonic functions in* $\mathbb{R}^2$ *subject to the Robin condition*, J. Math. Anal. Appl. **348** (2008), no. 2, 685–691. MR2445769

[16] S. R. Bell, *Quadrature domains and kernel function zipping*, Ark. Mat. **43** (2005), no. 2, 271–287. MR2173952

[17] S. R. Bell, P. Ebenfelt, D. Khavinson, and H. S. Shapiro, *On the classical Dirichlet problem in the plane with rational data*, J. Anal. Math. **100** (2006), 157–190. MR2303308

[18] S. R. Bell, P. Ebenfelt, D. Khavinson, and H. S. Shapiro, *Algebraicity in the Dirichlet problem in the plane with rational data*, Complex Var. Elliptic Equ. **52** (2007), no. 2-3, 235–244. MR2297773

[19] S. R. Bell, B. Gustafsson, and Z. A. Sylvan, *Szegő coordinates, quadrature domains, and double quadrature domains*, Comput. Methods Funct. Theory **11** (2011), no. 1, 25–44. MR2816943

[20] L. Bers, *An approximation theorem*, J. Analyse Math. **14** (1965), 1–4. MR0178287

[21] L. Bieberbach, *Analytische Fortsetzung* (German), Ergebnisse der Mathematik und ihrer Grenzgebiete (N.F.), Heft 3, Springer-Verlag, Berlin-Göttingen-Heidelberg, 1955. MR0068621

[22] J.-M. Bony and P. Schapira, *Existence et prolongement des solutions holomorphes des équations aux dérivées partielles* (French), Invent. Math. **17** (1972), 95–105. MR0338541

[23] M. Brelot and G. Choquet, *Polynômes harmoniques et polyharmoniques* (French), Second colloque sur les équations aux dérivées partielles, Bruxelles, 1954, Georges Thone, Liège; Masson & Cie, Paris, 1955, pp. 45–66. MR0069968

[24] H. C. Brinkman, *A calculation of the viscous force exerted by a flowing fluid on a dense swarm of particles*, Applied Scientific Research, 1 (1949), pp. 27–34.

[25] M. Chamberland and D. Siegel, *Polynomial solutions to Dirichlet problems*, Proc. Amer. Math. Soc. **129** (2001), no. 1, 211–217. MR1694451

[26] R. Courant and D. Hilbert, *Methods of mathematical physics. Vol. II: Partial differential equations*, (Vol. II by R. Courant.), Interscience Publishers (a division of John Wiley & Sons), New York-Lon don, 1962. MR0140802

[27] D. Crowdy, *A class of exact multipolar vortices*, Phys. Fluids **11** (1999), no. 9, 2556–2564. MR1707816

[28] D. Crowdy, *Quadrature domains and fluid dynamics*, Quadrature domains and their applications, Oper. Theory Adv. Appl., vol. 156, Birkhäuser, Basel, 2005, pp. 113–129. MR2129738

[29] D. G. Crowdy, *A note on viscous sintering and quadrature identities*, European J. Appl. Math. **10** (1999), no. 6, 623–634. Hele-Shaw flows and related problems (Oxford, 1998). MR1757945

[30] D. G. Crowdy and J. Roenby, *Hollow vortices, capillary water waves and double quadrature domains*, Fluid Dyn. Res. **46** (2014), no. 3, 031424, 12. MR3224254

[31] P. J. Davis, *The Schwarz function and its applications*, The Mathematical Association of America, Buffalo, N. Y., 1974. The Carus Mathematical Monographs, No. 17. MR0407252

[32] E. DiBenedetto and A. Friedman, *Bubble growth in porous media*, Indiana Univ. Math. J. **35** (1986), no. 3, 573–606. MR855175

[33] P. Dienes, *The Taylor series: an introduction to the theory of functions of a complex variable*, Dover Publications, Inc., New York, 1957. MR0089895

[34] P. Dive, *Attraction des ellipsoïdes homogènes et réciproques d'un théorème de Newton* (French), Bull. Soc. Math. France **59** (1931), 128–140. MR1504977

[35] J. Dunau, *Un problème de Cauchy caractéristique* (French), J. Math. Pures Appl. (9) **69** (1990), no. 3, 369–402. MR1070484

[36] P. L. Duren, *Theory of H^p spaces*, Pure and Applied Mathematics, Vol. 38, Academic Press, New York-London, 1970. MR0268655

[37] P. Ebenfelt, *Singularities encountered by the analytic continuation of solutions to Dirichlet's problem*, Complex Variables Theory Appl. **20** (1992), no. 1-4, 75–91. MR1284354

[38] P. Ebenfelt, *Holomorphic extension of solutions of elliptic partial differential equations and a complex Huygens' principle*, J. London Math. Soc. (2) **55** (1997), no. 1, 87–104. MR1423288

[39] P. Ebenfelt, B. Gustafsson, D. Khavinson, and M. Putinar, *Preface*, Quadrature domains and their applications, Oper. Theory Adv. Appl., vol. 156, Birkhäuser, Basel, 2005, pp. vii–x. MR2129732

[40] P. Ebenfelt and D. Khavinson, *On point to point reflection of harmonic functions across real-analytic hypersurfaces in* $\mathbf{R}^n$, J. Anal. Math. **68** (1996), 145–182. MR1403255

[41] P. Ebenfelt, D. Khavinson, and H. S. Shapiro, *Analytic continuation of Jacobi polynomial expansions*, Indag. Math. (N.S.) **8** (1997), no. 1, 19–31. MR1617790

[42] P. Ebenfelt, D. Khavinson, and H. S. Shapiro, *Extending solutions of holomorphic partial differential equations across real hypersurfaces*, J. London Math. Soc. (2) **57** (1998), no. 2, 411–432. MR1644225

[43] P. Ebenfelt, D. Khavinson, and H. S. Shapiro, *Algebraic aspects of the Dirichlet problem*, Quadrature domains and their applications, Oper. Theory Adv. Appl., vol. 156, Birkhäuser, Basel, 2005, pp. 151–172. MR2129740

[44] P. Ebenfelt and M. Viscardi, *On the solution of the Dirichlet problem with rational holomorphic boundary data*, Comput. Methods Funct. Theory **5** (2005), no. 2, 445–457. MR2205425
[45] A. Edelman and E. Kostlan, *How many zeros of a random polynomial are real?*, Bull. Amer. Math. Soc. (N.S.) **32** (1995), no. 1, 1–37. MR1290398
[46] V. M. Entov, P. I. Ètingof, and D. Ya. Kleinbock, *On nonlinear interface dynamics in Hele-Shaw flows*, European J. Appl. Math. **6** (1995), no. 5, 399–420. Complex analysis and free boundary problems (St. Petersburg, 1994). MR1363755
[47] A. Eremenko and E. Lundberg, *Non-algebraic quadrature domains*, Potential Anal. **38** (2013), no. 3, 787–804. MR3034600
[48] A. Eremenko and E. Lundberg, *Quasi-exceptional domains*, Pacific J. Math. **276** (2015), no. 1, 167–183. MR3366032
[49] P. I. Ètingof, *Integrability of filtration problems with a moving boundary* (Russian), Dokl. Akad. Nauk SSSR **313** (1990), no. 1, 42–47; English transl., Soviet Phys. Dokl. **35** (1990), no. 7, 625–628. MR1078920
[50] J. L. Walsh, *Recent Publications: Reviews: The Logarithmic Potential*, Amer. Math. Monthly **35** (1928), no. 5, 254–257. MR1521461
[51] M. N. Ferrers, *On the potentials of ellipsoids, ellipsoidal shells*, Quart. J. Pure and Appl. Math., 14 (1877), pp. 1–22.
[52] E. Fischer, *Über die Differentiationsprozesse der Algebra* (German), J. Reine Angew. Math. **148** (1918), 1–78. MR1580952
[53] A. Friedman, *Partial differential equations*, Holt, Rinehart and Winston, Inc., New York-Montreal, Que.-London, 1969. MR0445088
[54] A. Friedman and M. Sakai, *A characterization of null quadrature domains in* $\mathbf{R}^N$, Indiana Univ. Math. J. **35** (1986), no. 3, 607–610. MR855176
[55] D. Fuchs and S. Tabachnikov, *Mathematical omnibus*, American Mathematical Society, Providence, RI, 2007. Thirty lectures on classic mathematics. MR2350979
[56] P. R. Garabedian, *Partial differential equations with more than two independent variables in the complex domain*, J. Math. Mech. **9** (1960), 241–271. MR0120441
[57] P. R. Garabedian, *Lectures on function theory and partial differential equations*, Rice Univ. Studies **49** (1963), no. 4, 1–12. MR0176086
[58] P. R. Garabedian, *Partial differential equations*, John Wiley & Sons, Inc., New York-London-Sydney, 1964. MR0162045
[59] S. J. Gardiner and H. Render, *Harmonic functions which vanish on a cylindrical surface*, J. Math. Anal. Appl. **433** (2016), no. 2, 1870–1882. MR3398795
[60] S. J. Gardiner and H. Render, *A reflection result for harmonic functions which vanish on a cylindrical surface*, J. Math. Anal. Appl. **443** (2016), no. 1, 81–91. MR3508480
[61] L. Gårding, *Partial differential equations: Problems and uniformization in Cauchy's problem*, Lectures on Modern Mathematics, Vol. II, Wiley, New York, 1964, pp. 129–150. MR0178217
[62] L. Gårding, T. Kotake, and J. Leray, *Uniformisation et développement asymptotique de la solution du problème de Cauchy linéaire, à données holomorphes; analogie avec la théorie des ondes asymptotiques et approchées (Problème de Cauchy, I bis et VI)* (French), Bull. Soc. Math. France **92** (1964), 263–361. MR0196280
[63] I. M. Gel′fand and G. E. Shilov, *Generalized functions. Vol. 2. Spaces of fundamental and generalized functions*, Translated from the Russian by Morris D. Friedman, Amiel Feinstein and Christian P. Peltzer, Academic Press, New York-London, 1968. MR0230128
[64] R. P. Gilbert, *Function theoretic methods in partial differential equations*, Mathematics in Science and Engineering, Vol. 54, Academic Press, New York-London, 1969. MR0241789
[65] A. B. Givental, *Polynomiality of electrostatic potentials*, Uspekhi Mat. Nauk, 39 (1984), pp. 253–254.
[66] P. Griffiths and J. Harris, *Principles of algebraic geometry*, Wiley-Interscience [John Wiley & Sons], New York, 1978. Pure and Applied Mathematics. MR507725
[67] B. Gustafsson, *Quadrature identities and the Schottky double*, Acta Appl. Math. **1** (1983), no. 3, 209–240. MR726725
[68] B. Gustafsson, *Application of half-order differentials on Riemann surfaces to quadrature identities for arc-length*, J. Analyse Math. **49** (1987), 54–89. MR928507

[69] B. Gustafsson and M. Putinar, *Hyponormal Quantization of Planar Domains: Exponential Transform in Dimension Two*, Lecture Notes in Mathematics, Birkhäuser/Springer, Cham, 2017.

[70] B. Gustafsson and H. S. Shapiro, *What is a quadrature domain?*, Quadrature domains and their applications, Oper. Theory Adv. Appl., vol. 156, Birkhäuser, Basel, 2005, pp. 1–25. MR2129734

[71] J. Hadamard, *Le problème de Cauchy et les équations aux dérivées partielles linéaires hyperboliques: lecons professées à l'université Yale* (French), Hermann et Cie, Paris, édition revue et notablement augmentée ed., 1932.

[72] J. Hadamard, *Lectures on Cauchy's problem in linear partial differential equations*, Dover Publications, New York, 1953. MR0051411

[73] Y. Hamada, *Les singularités des solutions du problème de Cauchy à données holomorphes* (French, with English summary), Recent developments in hyperbolic equations (Pisa, 1987), Pitman Res. Notes Math. Ser., vol. 183, Longman Sci. Tech., Harlow, 1988, pp. 82–95. MR984361

[74] Y. Hamada, J. Leray, and C. Wagschal, *Systèmes d'équations aux dérivées partielles à caractéristiques multiples: problème de Cauchy ramifié; hyperbolicité partielle* (French), J. Math. Pures Appl. (9) **55** (1976), no. 3, 297–352. MR0435614

[75] L. J. Hansen and H. S. Shapiro, *Functional equations and harmonic extensions*, Complex Variables Theory Appl. **24** (1994), no. 1-2, 121–129. MR1269837

[76] L. Hauswirth, F. Hélein, and F. Pacard, *On an overdetermined elliptic problem*, Pacific J. Math. **250** (2011), no. 2, 319–334. MR2794602

[77] W. K. Hayman, *Power series expansions for harmonic functions*, Bull. London Math. Soc. **2** (1970), 152–158. MR0267114

[78] W. K. Hayman and B. Korenblum, *Representation and uniqueness theorems for polyharmonic functions*, J. Anal. Math. **60** (1993), 113–133. MR1253232

[79] J. W. Helton, S. McCullough, M. Putinar, and V. Vinnikov, *Convex matrix inequalities versus linear matrix inequalities*, IEEE Trans. Automat. Control **54** (2009), no. 5, 952–964. MR2518108

[80] J. W. Helton and V. Vinnikov, *Linear matrix inequality representation of sets*, Comm. Pure Appl. Math. **60** (2007), no. 5, 654–674. MR2292953

[81] P. Henrici, *A survey of I. N. Vekua's theory of elliptic partial differential equations with analytic coefficients*, Z. Angew. Math. Phys. **8** (1957), 169–203. MR0085427

[82] P. Henrici, *Applied and computational complex analysis. Vol. 3*, Pure and Applied Mathematics (New York), John Wiley & Sons, Inc., New York, 1986. Discrete Fourier analysis—Cauchy integrals—construction of conformal maps—univalent functions; A Wiley-Interscience Publication. MR822470

[83] G. Herglotz, *Gesammelte Schriften* (German), Vandenhoeck & Ruprecht, Göttingen, 1979. With introductory articles by Peter Bergmann, S. S. Chern, Ronald B. Guenther, Claus Müller, Theodor Schneider and H. Wittich; Edited and with a foreword by Hans Schwerdtfeger. MR526569

[84] E. Hölder, *Über eine potentialtheoretische Eigenschaft der Ellipse* (German), Math. Z. **35** (1932), no. 1, 632–643. MR1545322

[85] L. Hörmander, *Linear partial differential operators*, Die Grundlehren der mathematischen Wissenschaften, Bd. 116, Academic Press, Inc., Publishers, New York; Springer-Verlag, Berlin-Göttingen-Heidelberg, 1963. MR0161012

[86] L. Hörmander, *The analysis of linear partial differential operators. I*, Grundlehren der Mathematischen Wissenschaften [Fundamental Principles of Mathematical Sciences], vol. 256, Springer-Verlag, Berlin, 1983. Distribution theory and Fourier analysis. MR717035

[87] L. Hörmander, *An introduction to complex analysis in several variables*, 3rd ed., North-Holland Mathematical Library, vol. 7, North-Holland Publishing Co., Amsterdam, 1990. MR1045639

[88] S. D. Howison, *Cusp development in Hele-Shaw flow with a free surface*, SIAM J. Appl. Math. **46** (1986), no. 1, 20–26. MR821438

[89] K. Igari, *Propagation of singularities in the ramified Cauchy problem for a class of operators with non-involutive multiple characteristics*, Publ. Res. Inst. Math. Sci. **35** (1999), no. 2, 285–307. MR1684207

[90] E. L. Ince, *Ordinary Differential Equations*, Dover Publications, New York, 1944. MR0010757

[91] V. K. Ivanov, *The inverse problem of potential for a body closely approximated by a given body* (Russian), Izv. Akad. Nauk SSSR. Ser. Mat. **20** (1956), 793–818. MR0084586

[92] F. John, *On linear partial differential equations with analytic coefficients. Unique continuation of data*, Comm. Pure Appl. Math. **2** (1949), 209–253. MR0036930

[93] F. John, *Partial differential equations*, 2nd ed., Springer-Verlag, New York-Heidelberg, 1975. Applied mathematical sciences, Vol. 1. MR0350162

[94] G. Johnsson, *The Cauchy problem in* $\mathbf{C}^N$ *for linear second order partial differential equations with data on a quadric surface*, Trans. Amer. Math. Soc. **344** (1994), no. 1, 1–48. MR1242782

[95] G. Johnsson, *Global existence results for linear analytic partial differential equations*, J. Differential Equations **115** (1995), no. 2, 416–440. MR1310939

[96] L. Karp, *Construction of quadrature domains in* $\mathbf{R}^n$ *from quadrature domains in* $\mathbf{R}^2$, Complex Variables Theory Appl. **17** (1992), no. 3-4, 179–188. MR1147048

[97] L. Karp, *On the Newtonian potential of ellipsoids*, Complex Variables Theory Appl. **25** (1994), no. 4, 367–371. MR1314951

[98] M. Kashiwara, T. Monteiro Fernandes, and P. Schapira, *Truncated microsupport and holomorphic solutions of D-modules* (English, with English and French summaries), Ann. Sci. École Norm. Sup. (4) **36** (2003), no. 4, 583–599. MR2013927

[99] O. D. Kellogg, *Foundations of potential theory*, Reprint from the first edition of 1929. Die Grundlehren der Mathematischen Wissenschaften, Band 31, Springer-Verlag, Berlin-New York, 1967. MR0222317

[100] V. P. Khavin, *A remark on Taylor series of harmonic functions* (Russian), Multidimensional complex analysis (Russian), Akad. Nauk SSSR Sibirsk. Otdel., Inst. Fiz., Krasnoyarsk, 1985, pp. 192–197, 277. MR899349

[101] D. Khavinson, *On reflection of harmonic functions in surfaces of revolution*, Complex Variables Theory Appl. **17** (1991), no. 1-2, 7–14. MR1123797

[102] D. Khavinson, *Singularities of harmonic functions in* $\mathbf{C}^n$, Several complex variables and complex geometry, Part 3 (Santa Cruz, CA, 1989), Proc. Sympos. Pure Math., vol. 52, Amer. Math. Soc., Providence, RI, 1991, pp. 207–217. MR1128595

[103] D. Khavinson, *Holomorphic partial differential equations and classical potential theory*, Universidad de La Laguna, Departamento de Análisis Matemático, La Laguna, 1996. MR1392698

[104] D. Khavinson, *Cauchy's problem for harmonic functions with entire data on a sphere*, Canad. Math. Bull. **40** (1997), no. 1, 60–66. MR1443726

[105] D. Khavinson and E. Lundberg, *The search for singularities of solutions to the Dirichlet problem: recent developments*, Hilbert spaces of analytic functions, CRM Proc. Lecture Notes, vol. 51, Amer. Math. Soc., Providence, RI, 2010, pp. 121–132. MR2648870

[106] D. Khavinson and E. Lundberg, *A tale of ellipsoids in potential theory*, Notices Amer. Math. Soc. **61** (2014), no. 2, 148–156. MR3156681

[107] D. Khavinson, E. Lundberg, and R. Teodorescu, *An overdetermined problem in potential theory*, Pacific J. Math. **265** (2013), no. 1, 85–111. MR3095114

[108] D. Khavinson, M. Mineev-Weinstein, and M. Putinar, *Planar elliptic growth*, Complex Anal. Oper. Theory **3** (2009), no. 2, 425–451. MR2504763

[109] D. Khavinson, M. Mineev-Weinstein, M. Putinar, and R. Teodorescu, *Lemniscates do not survive Laplacian growth*, Math. Res. Lett. **17** (2010), no. 2, 335–341. MR2644380

[110] D. Khavinson and H. S. Shapiro, *The Schwarz potential in* $\mathbb{R}^n$ *and Cauchy's problem for the Laplace equation*, Tech. Rep. TRITA-MAT-1989-36, Royal Institute of Technology, Stockholm, 1989, 112 pp., preprint available at `http://shell.cas.usf.edu/~dkhavins/publications.html`.

[111] D. Khavinson and H. S. Shapiro, *Remarks on the reflection principle for harmonic functions*, J. Analyse Math. **54** (1990), 60–76. MR1041175

[112] D. Khavinson and H. S. Shapiro, *The Vekua hull of a plane domain*, Complex Variables Theory Appl. **14** (1990), no. 1-4, 117–128. MR1048711

[113] D. Khavinson and H. S. Shapiro, *Dirichlet's problem when the data is an entire function*, Bull. London Math. Soc. **24** (1992), no. 5, 456–468. MR1173942

[114] D. Khavinson and H. S. Shapiro, *The heat equation and analytic continuation: Ivar Fredholm's first paper*, Exposition. Math. **12** (1994), no. 1, 79–95. MR1267629

[115] D. Khavinson and H. S. Shapiro, *On a uniqueness property of harmonic functions*, Comput. Methods Funct. Theory **8** (2008), no. 1-2, 143–150. MR2419467

[116] D. Khavinson and N. Stylianopoulos, *Recurrence relations for orthogonal polynomials and algebraicity of solutions of the Dirichlet problem*, Around the research of Vladimir Maz'ya. II, Int. Math. Ser. (N. Y.), vol. 12, Springer, New York, 2010, pp. 219–228. MR2676175

[117] S. Ya. Khavinson, *Best approximation by linear superpositions (approximate nomography)*, Translations of Mathematical Monographs, vol. 159, American Mathematical Society, Providence, RI, 1997. Translated from the Russian manuscript by D. Khavinson. MR1421322

[118] C. O. Kiselman, *Prolongement des solutions d'une équation aux dérivées partielles à coefficients constants* (French), Bull. Soc. Math. France **97** (1969), 329–356. MR0267259

[119] S. v. Kowalevsky, *Zur Theorie der partiellen Differentialgleichung* (German), J. Reine Angew. Math. **80** (1875), 1–32. MR1579652

[120] N. S. Landkof, *Foundations of modern potential theory*, Springer-Verlag, New York-Heidelberg, 1972. Translated from the Russian by A. P. Doohovskoy; Die Grundlehren der mathematischen Wissenschaften, Band 180. MR0350027

[121] S. Lang, *Algebra*, 3rd ed., Graduate Texts in Mathematics, vol. 211, Springer-Verlag, New York, 2002. MR1878556

[122] P. D. Lax, *Differential equations, difference equations and matrix theory*, Comm. Pure Appl. Math. **11** (1958), 175–194. MR0098110

[123] L. Lempert, *Recursion for orthogonal polynomials on complex domains*, Fourier analysis and approximation theory (Proc. Colloq., Budapest, 1976), Vol. II, Colloq. Math. Soc. János Bolyai, vol. 19, North-Holland, Amsterdam-New York, 1978, pp. 481–494. MR540325

[124] J. Leray, *Le problème de Cauchy pour une équation linéaire à coefficients polynomiaux* (French), C. R. Acad. Sci. Paris **242** (1956), 953–959. MR0077778

[125] J. Leray, *Problème de Cauchy. I. Uniformisation de la solution du problème linéaire analytique de Cauchy près de la variété qui porte les données de Cauchy* (French), Bull. Soc. Math. France **85** (1957), 389–429. MR0103328

[126] J. Leray, *Uniformisation de la solution du problème linéaire analytique de Cauchy près de la variété qui porte les données de Cauchy* (French), C. R. Acad. Sci. Paris **245** (1957), 1483–1488. MR0093634

[127] J. Leray, *La solution unitaire d'un opérateur différentiel linéaire (Problème de Cauchy. II.)* (French), Bull. Soc. Math. France **86** (1958), 75–96. MR0121573

[128] J. Leray, *Le calcul différentiel et intégral sur une variété analytique complexe. (Problème de Cauchy. III)* (French), Bull. Soc. Math. France **87** (1959), 81–180. MR0125984

[129] J. Leray, *Un prolongementa de la transformation de Laplace qui transforme la solution unitaires d'un opérateur hyperbolique en sa solution élémentaire. (Problème de Cauchy. IV.)* (French), Bull. Soc. Math. France **90** (1962), 39–156. MR0144077

[130] A. S. Lewis, P. A. Parrilo, and M. V. Ramana, *The Lax conjecture is true*, Proc. Amer. Math. Soc. **133** (2005), no. 9, 2495–2499. MR2146191

[131] H. Lewy, *On the reflection laws of second order differential equations in two independent variables*, Bull. Amer. Math. Soc. **65** (1959), 37–58. MR0104048

[132] S. G. Llewellyn Smith and D. G. Crowdy, *Structure and stability of hollow vortex equilibria*, J. Fluid Mech. **691** (2012), 178–200. MR2872976

[133] J. E. C. Lope and H. Tahara, *On the analytic continuation of solutions to nonlinear partial differential equations*, J. Math. Pures Appl. (9) **81** (2002), no. 9, 811–826. MR1940368

[134] I. Loutsenko and O. Yermolayeva, *Non-Laplacian growth: exact results*, Phys. D **235** (2007), no. 1-2, 56–61. MR2385116

[135] E. Lundberg, *Dirichlet's problem and complex lightning bolts*, Comput. Methods Funct. Theory **9** (2009), no. 1, 111–125. MR2478266

[136] E. Lundberg, *Laplacian growth, elliptic growth, and singularities of the Schwarz potential*, J. Phys. A **44** (2011), no. 13, 135202, 20. MR2776207

[137] E. Lundberg and H. Render, *The Khavinson-Shapiro conjecture and polynomial decompositions*, J. Math. Anal. Appl. **376** (2011), no. 2, 506–513. MR2747774

[138] E. Lundberg and V. Totik, *Lemniscate growth*, Anal. Math. Phys. **3** (2013), no. 1, 45–62. MR3015630

[139] W. D. MacMillan, *The theory of the potential*, MacMillan's Theoretical Mechanics, Dover Publications, Inc., New York, 1958. MR0100172

[140] J. E. Mapel-Hemmati, *A Fuchs-type theorem for partial differential equations*, J. Math. Anal. Appl. **239** (1999), no. 1, 30–37. MR1719092

[141] J. E. McCarthy and L. Yang, *Subnormal operators and quadrature domains*, Adv. Math. **127** (1997), no. 1, 52–72. MR1445362

[142] P. McCoy, *Polynomial expansions of analytic functions by function-theoretic methods*, Partial differential equations with complex analysis, Pitman Res. Notes Math. Ser., vol. 262, Longman Sci. Tech., Harlow, 1992, pp. 122–133. MR1197609

[143] M. Miyake, *Global and local Goursat problems in a class of holomorphic or partially holomorphic functions*, J. Differential Equations **39** (1981), no. 3, 445–463. MR612597

[144] N. Nadirashvili, *Private communication with H. S. Shapiro*, (1994).

[145] Z. Nehari, *On the singularities of Legendre expansions*, J. Rational Mech. Anal. **5** (1956), 987–992. MR0080747

[146] Y. Nesterov and A. Nemirovskii, *Interior-point polynomial algorithms in convex programming*, SIAM Studies in Applied Mathematics, vol. 13, Society for Industrial and Applied Mathematics (SIAM), Philadelphia, PA, 1994. MR1258086

[147] D. J. Newman and H. S. Shapiro, *A Hilbert space of entire functions related to the operational calculus*. mimeographed notes, 1964.

[148] D. J. Newman and H. S. Shapiro, *Fischer spaces of entire functions*, in Entire Functions and Related Parts of Analysis (Proc. Sympos. Pure Math., La Jolla, Calif., 1966), Amer. Math. Soc., Providence, R.I., 1968, pp. 360–369.

[149] W. Nikliborc, *Eine Bemerkung über die Volumpotentiale* (German), Math. Z. **35** (1932), no. 1, 625–631. MR1545320

[150] J. Persson, *On the local and global non-characteristic Cauchy problem when the solutions are holomorphic functions or analytic functionals or analytic functionals in the space variables*, Ark. Mat. **9** (1971), 171–180 (1971). MR0318702

[151] I. G. Petrovsky, *Lectures on partial differential equations*, Dover Publications, Inc., New York, 1991. Translated from the Russian by A. Shenitzer; Reprint of the 1964 English translation. MR1160355

[152] M. Putinar and N. S. Stylianopoulos, *Finite-term relations for planar orthogonal polynomials*, Complex Anal. Oper. Theory **1** (2007), no. 3, 447–456. MR2336033

[153] T. Ransford, *Potential theory in the complex plane*, London Mathematical Society Student Texts, vol. 28, Cambridge University Press, Cambridge, 1995. MR1334766

[154] C. Reid, *Hilbert*, Copernicus, New York, 1996. Reprint of the 1970 original. MR1391242

[155] H. Render, *Real Bargmann spaces, Fischer decompositions, and sets of uniqueness for polyharmonic functions*, Duke Math. J. **142** (2008), no. 2, 313–352. MR2401623

[156] H. Render, *Cauchy, Goursat and Dirichlet problems for holomorphic partial differential equations*, Comput. Methods Funct. Theory **10** (2010), no. 2, 519–554. MR2791323

[157] H. Render, *Harmonic divisors and rationality of zeros of Jacobi polynomials*, Ramanujan J. **31** (2013), no. 3, 257–270. MR3081667

[158] H. Render, *A characterization of the Khavinson-Shapiro conjecture via Fischer operators*, Potential Anal. **45** (2016), no. 3, 539–543. MR3554402

[159] R. Restrepo López, *On reflection principles supported on a finite set*, J. Math. Anal. Appl. **351** (2009), no. 2, 556–566. MR2473961

[160] S. Richardson, *Hele-shaw flows with a free boundary produced by the injection of fluid into a narrow channel*, J. Fluid Mech., 56 (1972), pp. 609–618.

[161] W. Rudin, *Functional analysis*, McGraw-Hill Book Co., New York-Düsseldorf-Johannesburg, 1973. McGraw-Hill Series in Higher Mathematics. MR0365062

[162] W. Rudin, *Principles of mathematical analysis*, 3rd ed., McGraw-Hill Book Co., New York-Auckland-Düsseldorf, 1976. International Series in Pure and Applied Mathematics. MR0385023

[163] E. B. Saff and V. Totik, *Logarithmic potentials with external fields*, Grundlehren der Mathematischen Wissenschaften [Fundamental Principles of Mathematical Sciences], vol. 316, Springer-Verlag, Berlin, 1997. Appendix B by Thomas Bloom. MR1485778

[164] M. Sakai, *Null quadrature domains*, J. Analyse Math. **40** (1981), 144–154 (1982). MR659788

[165] M. Sakai, *Quadrature domains*, Lecture Notes in Mathematics, vol. 934, Springer-Verlag, Berlin-New York, 1982. MR663007

[166] A. Savin and B. Sternin, *Introduction to complex theory of differential equations*, Frontiers in Mathematics, Birkhäuser/Springer, Cham, 2017. MR3643052

[167] T. V. Savina, *On the dependence of the reflection operator on boundary conditions for biharmonic functions*, J. Math. Anal. Appl. **370** (2010), no. 2, 716–725. MR2651690

[168] T. Savina, *On non-local reflection for elliptic equations of the second order in* $\mathbb{R}^2$ *(the Dirichlet condition)*, Trans. Amer. Math. Soc. **364** (2012), no. 5, 2443–2460. MR2888214

[169] T. V. Savina, B. Yu. Sternin, and V. E. Shatalov, *On the reflection law for the Helmholtz equation* (Russian), Dokl. Akad. Nauk SSSR **322** (1992), no. 1, 48–51; English transl., Soviet Math. Dokl. **45** (1992), no. 1, 42–45. MR1158949

[170] P. Schapira, *Hyperbolic systems and propagation on causal manifolds*, Lett. Math. Phys. **103** (2013), no. 10, 1149–1164. MR3077960

[171] B. V. Shabat, *Introduction to complex analysis. Part II*, Translations of Mathematical Monographs, vol. 110, American Mathematical Society, Providence, RI, 1992. Functions of several variables; Translated from the third (1985) Russian edition by J. S. Joel. MR1192135

[172] H. Shahgholian, *On the Newtonian potential of a heterogeneous ellipsoid*, SIAM J. Math. Anal. **22** (1991), no. 5, 1246–1255. MR1112506

[173] A. D. Vaĭnshteĭn and B. Z. Shapiro, *Multidimensional analogues of the Newton and Ivory theorems* (Russian), Funktsional. Anal. i Prilozhen. **19** (1985), no. 1, 20–24, 96. MR783702

[174] H. S. Shapiro, *Unbounded quadrature domains*, Complex analysis, I (College Park, Md., 1985), Lecture Notes in Math., vol. 1275, Springer, Berlin, 1987, pp. 287–331. MR922309

[175] H. S. Shapiro, *An algebraic theorem of E. Fischer, and the holomorphic Goursat problem*, Bull. London Math. Soc. **21** (1989), no. 6, 513–537. MR1018198

[176] H. S. Shapiro, *The Schwarz function and its generalization to higher dimensions*, University of Arkansas Lecture Notes in the Mathematical Sciences, vol. 9, John Wiley & Sons, Inc., New York, 1992. A Wiley-Interscience Publication. MR1160990

[177] M. Shub and S. Smale, *Complexity of Bezout's theorem. II. Volumes and probabilities*, Computational algebraic geometry (Nice, 1992), Progr. Math., vol. 109, Birkhäuser Boston, Boston, MA, 1993, pp. 267–285. MR1230872

[178] J. Siciak, *Holomorphic continuation of harmonic functions*, Ann. Polon. Math. **29** (1974), 67–73. Collection of articles dedicated to the memory of Tadeusz Ważewski. MR0352530

[179] L. Sretensky, *On the inverse problem of potential theory* (Russian), Izv. Akad. Nauk SSSR Ser. Mat., 2 (1938), pp. 551–570.

[180] E. M. Stein and G. Weiss, *Introduction to Fourier analysis on Euclidean spaces*, Princeton University Press, Princeton, N.J., 1971. Princeton Mathematical Series, No. 32. MR0304972

[181] B. Sternin and V. Shatalov, *Differential equations on complex manifolds*, Mathematics and its Applications, vol. 276, Kluwer Academic Publishers Group, Dordrecht, 1994. MR1391964

[182] B. Yu. Sternin, Kh. S. Shapiro, and V. E. Shatalov, *Schwarz potential and singularities of the solutions of the branching Cauchy problem* (Russian), Dokl. Akad. Nauk **341** (1995), no. 3, 307–309. MR1370569

[183] D. W. Stroock, *Weyl's lemma, one of many*, Groups and analysis, London Math. Soc. Lecture Note Ser., vol. 354, Cambridge Univ. Press, Cambridge, 2008, pp. 164–173. MR2528466

[184] E. Study, *Einige elementare Bemerkungen über den Prozeß der analytischen Fortsetzung* (German), Math. Ann. **63** (1906), no. 2, 239–245. MR1511403

[185] N. Stylianopoulos, *Strong asymptotics for Bergman polynomials over domains with corners and applications*, Constr. Approx. **38** (2013), no. 1, 59–100. MR3078274

[186] G. Szegő, *On the singularities of zonal harmonic expansions*, J. Rational Mech. Anal. **3** (1954), 561–564. MR0062880

[187] G. Szegő, *Orthogonal polynomials*, American Mathematical Society Colloquium Publications, Vol. 23. Revised ed, American Mathematical Society, Providence, R.I., 1959. MR0106295

[188] S. Tanveer, *Surprises in viscous fingering*, J. Fluid Mech. **409** (2000), 273–308. MR1756392

[189] F.-R. Tian, *Hele-Shaw problems in multidimensional spaces*, J. Nonlinear Sci. **10** (2000), no. 2, 275–290. MR1743401

[190] V. Totik, *Bernstein-type inequalities*, J. Approx. Theory **164** (2012), no. 10, 1390–1401. MR2961187

[191] M. Traizet, *Classification of the solutions to an overdetermined elliptic problem in the plane*, Geom. Funct. Anal. **24** (2014), no. 2, 690–720. MR3192039

[192] M. Traizet, *Hollow vortices and minimal surfaces*, J. Math. Phys. **56** (2015), no. 8, 083101, 18. MR3455335

[193] Y. Tsuno, *On the prolongation of local holomorphic solutions of partial differential equations*, J. Math. Soc. Japan **26** (1974), 523–548. MR0361389

[194] A. Tychonoff, *Théoremes d'unicité pour l'équation de la chaleur* (French), Rec. Math. Moscou, 42 (1935), pp. 199–215.

[195] A. N. Varchenko and P. I. Ètingof, *Why the boundary of a round drop becomes a curve of order four*, University Lecture Series, vol. 3, American Mathematical Society, Providence, RI, 1992. MR1190012

[196] I. N. Vekua, *New methods for solving elliptic equations*, Translated from the Russian by D. E. Brown. Translation edited by A. B. Tayler, North-Holland Publishing Co., Amsterdam; Interscience Publishers John Wiley & Sons, Inc., New York, 1967. North-Holland Series in Applied Mathematics and Mechanics, Vol. 1. MR0212370

[197] C. Vinzant, *What is ... a spectrahedron?*, Notices Amer. Math. Soc. **61** (2014), no. 5, 492–494. MR3203240

[198] C. Wagschal, *Sur le problème de Cauchy ramifié* (French), J. Math. Pures Appl. (9) **53** (1974), 147–163. MR0382832

[199] G. G. Walter and A. J. Zayed, *Real singularities of singular Sturm-Liouville expansions*, SIAM J. Math. Anal. **18** (1987), no. 1, 219–227. MR871833

[200] G. N. Watson, *A Treatise on the Theory of Bessel Functions*, Cambridge University Press, Cambridge, England; The Macmillan Company, New York, 1944. MR0010746

[201] A. Zayed, M. Freund, and E. Görlich, *A theorem of Nehari revisited*, Complex Variables Theory Appl. **10** (1988), no. 1, 11–22. MR946095

[202] A. I. Zayed, *On the singularities of the continuous Jacobi transform when $\alpha + \beta = 0$*, Proc. Amer. Math. Soc. **101** (1987), no. 1, 67–75. MR897072

[203] M. Zerner, *Domaines d'holomorphie des fonctions vérifiant une équation aux dérivées partielles* (French), C. R. Acad. Sci. Paris Sér. A-B **272** (1971), A1646–A1648. MR0433016

Index